11/03

INNOVATION AND THE COMMUNICATIONS REVOLUTION

from the Victorian pioneers to broadband Internet

John Bray

The Institution of Electrical Engineers

Published by: The Institution of Electrical Engineers, London,
United Kingdom

British Library Cataloguing in Publication Data

Bray, J.
Innovation and the communications revolution: from the
Victorian pioneers to broadband Internet
1. Telecommunication – History 2. Telecommunication –
Technological innovations
I. Title II. Institution of Electrical Engineers
621.3'82

ISBN 0 85296 218 5

Typeset by RefineCatch Ltd, Bungay, Suffolk
Printed in England by MPG Books Ltd., Bodmin, Cornwall

The good effects wrought by founders of cities, law givers, fathers of the peoples, extirpators of tyrants and heroes of that class extend but for short times, whereas the work of the inventor, though a thing of less pomp and show, is felt everywhere and lasts for ever.

Francis Bacon, 1561–1626

Contents

List of figures

Foreword

It has been said that the 'electronic revolution' of the past half-century constitutes the most important event in human history since the birth of Jesus Christ. Certainly its impact upon our daily lives and, indeed, its predication of our future, are so total as to be unquantifiable.

However, the subject of this book pre-dates, encompasses and looks beyond the discovery of the semi-conductor and all its allied technological achievements. Not only that, its author has also chosen to present his account of the 'dramatic advances in telecommunication and broadcasting that have occurred in the last century and a half' as an essentially human story. He writes not only about the achievements themselves, but also about those who themselves made those giant strides a reality. It would be hard to imagine any candidate more qualified to undertake a task as monumental in scale as importance – he was himself a principal player in the subject of his narrative. John Bray, already a world figure in telecommunications, proves himself in this book to be a splendid writer.

We first met in July 1962 at the British Post Office satellite communication Earth station at Goonhilly, Cornwall when I was privileged to present to the world a commentary on the first ever television transmission from Europe to America via the Telstar satellite and the ATT/Bell Earth station at Andover, Maine.

I am grateful for the chances of my professional life to have been thrown into close contact – albeit usually all too briefly and often under considerable pressure – with some of the great figures of our time in the realms of science and technology. That one of them should invite me to write a foreword to his work is a compliment of which I am deeply appreciative.

John Bray expresses the hope that this work may inspire young people to follow the path he has trodden with such distinction. I cannot wait to present it to my grandchildren.

Raymond Baxter
Broadcaster and Former Presenter of *Tomorrow's World*

Preface

This book is an account of the dramatic advances in telecommunications, broadcasting and information technology that have occurred in the last century and a half, told through the stories of the people who made it happen. It brings together the lives and achievements of the men whose innovative and creative work in science, engineering and mathematics from the mid-1800s onwards made possible present-day national and international telecommunications services, sound and television broadcasting, and the World Wide Web.

These major developments can, in many cases, be traced back to specific advances in scientific knowledge, novel inventions or new system concepts created by certain individuals, whose work can now be seen to have provided the essential keys to progress.

The book explores the background and motivation of each of these men and the social and economic environment in which they worked. It describes what they felt about their discoveries and the significance of their contributions to the modern world.

Few of the technological advances described arose from sponsored, market-orientated cost-effectiveness studies; they were more often the result of original creative thought – and sometimes intuition. These were individuals who foresaw a communication need and the scientific and engineering possibilities for meeting it.

The book is a history of technological innovators and innovations in telecommunications, broadcasting and information access and processing that are taken for granted in the modern world but which will have a profound effect on the way people live and work in the future.

Out of the blue?

Two books that were part of the writer's reading in 'light current' electrical engineering at Imperial College, London in the 1930s were Rollo Appleyard's *Pioneers of Electrical Communication*, published in 1930, and E.T.Bell's *Men of Mathematics*, published in 1937. By their approach these books made the reader aware of the reality of the innovative process, and made it clear that radically new ideas and technological advances do not appear 'out of the blue'.

More often than not these ideas are the product of concentrated and continuing thought in a chosen field, allied to intuition, and leading to a quantum jump forward in understanding. They led at least one reader to pursue a professional career as near as possible to the 'sharp cutting edge' of technological advance in telecommunications.[1]

Since the 1930s the development of telecomunications and broadcasting has continued at an accelerating pace, with many innovative and clearly defined steps along the way. In many cases the pioneers responsible can readily be identified – generally by consensus judgement, for example by scientific and engineering institutions. Other cases may be less obvious and, in a commercially competitive world, may need to be selected for the purposes of this book on a more personal and subjective basis, for which the writer makes apology in advance where less than justice may have been done to other claimants for pioneer status.

Clearly it is not possible in a single book covering such a broad field to follow in detail the ongoing development beyond the first highly innovative step, but an endeavour has been made in each case to indicate in broad terms why the selection has been made and justified in terms of its long-term impact.

Innovation in telecommunication and broadcasting technology is not a national monopoly – the ferment of new ideas exists in many organizations and in many countries, and sometimes in the minds of lone individuals. While in this book some prominence is given to British innovators and inventors, it is clear that many key ideas originated elsewhere, notably in the USA, Germany, France and Austria, and more recently in Japan.

Engineers, scientists and mathematicians in Bell Telephone Laboratories, USA have made many important contributions to the advance of telecommunications, and it has been the writer's privilege to have been associated with some of these, from the development of transatlantic short-wave radio communication in the 1930s to co-operation in the pioneering Telstar communication satellite project of the 1960s. These contacts were enhanced and given a more personal basis by a Commonwealth Fund (Harkness Foundation) Fellowship spent at the Bell Laboratories at Murray Hill and Holmdel during 1955–1956.

Aims of the book

It is hoped that this book will provide a human thread through the history of technological advances in telecommunications and broadcasting that will inspire school-leavers and those studying in colleges and universities to seek a rewarding career in one of the most important developments of our time – one in which there is much to be achieved.

[1] In the writer's case, as Director of Research of the Post Office Engineering Dept., now British Telecom, from 1966 to 1975. See autobiography *Then, Now and Tomorrow* (Book Guild, 1999).

The book is also aimed at the innovators of tomorrow in the hope that the innovators of yesterday provide inspiration to them to match the communication needs of the future.

No apology is made for concentrating on technological innovation; without the supporting technology none of the far-reaching services now available in telecommunications and broadcasting would have been possible. Nevertheless, it is clear that innovation in design, manufacture, finance and exploitation also have important roles to play in bringing the results of technological innovation to fruition, and these too create new career possibilities.

Acknowledgements

The writer's thanks are due to authors, editors and publishers of the many books which have been found useful in preparing the text, and to which reference has been made in the bibliography at the end of each chapter.

Certain books, notably W.J. Baker's *History of the Marconi Company*, the *History of Science and Engineering in the Bell System: Transmission Technology (1925–1975)*, edited by Dr Gene O'Neill, and George Brown's *A Part of Which I Was*: Recollections of a Research Engineer (which gives a fascinating insight into the development of colour television in the USA) have been particularly useful. Personal records of Bell Laboratories people such as H.T. Friis's *75 Years in an Exciting World*, J.R. Pierce's *The Beginnings of Satellite Communication*, and R. Kompfner's *The Invention of the Travelling-Wave Tube* (which had an important role in microwave radio-relay and satellite systems) have been valuable.

The publication *Bell Laboratories Record* has provided a very readable account of developments in that organisation. In the UK the *Post Office Electrical Engineers' Journal*, the *Journal of British Telecommunications Engineering* and the *Proceedings of the Institution of Electrical Engineers*, London have provided valuable documentation.

The International Telecommunication Union publication *From Semaphore to Satellite*, published on its Centenary Anniversary in 1965, has provided a useful picture of telecommunications development to that date. In the specialised exchange switching field Robert Chappuis' and Amos Joel's *100 Years of Telephone Switching: 1878–1978* gives a comprehensive account for that period.

The author's thanks are due to former colleagues in the Engineering Department of the British Post Office (now British Telecom), including Charles Hughes, Michael Holmes, Roy Harris, J. Martin, M.B. Williams and Murray Laver for their always helpful advice and to Dr J.R. Tillman, former deputy director of research PO/BT, for material for Chapter 12 on the invention of the transistor. Professor Gerry White has been particularly helpful for advice on current developments, and Professor D. Cheeseman has updated Chapter 21 on developments in mobile radio communication.

The author is particularly grateful to Dr Gene O'Neill, formerly of Bell Laboratories/ATT Co., for encouragement during the writing of the book and constructive reviewing of the text.

Various firms and organisations, including the Marconi Co., Standard Telephones and Cables Ltd, Corning Glass USA, ATT Co.mpany (ATT Co.) and Bell Laboratories, USA, the Royal Television Society, UK, the David Sarnoff Research Centre Archives, USA, the BBC and the Independent Broadcasting Authority have provided copies of documents and photographs. The US National Academy of Engineering has supplied information on the Draper Award made to the co-inventors of the microchip. The Library of the Institution of Electrical Engineers, UK has been particularly helpful in providing loan copies of relevant books.

Several organisations have provided photographs and illustrations of historic and technical interest, including:

Hunterian Art Gallery, University of Glasgow (Lord Kelvin)
National Portrait Gallery (Lord Rutherford)
IEE, London (Sir J.J. Thomson)
US Library of Congress (Alexander Graham Bell)
Marconi Co. Archives (Marconi and Writtle Transmitter)
BBC (Capt. P.P. Eckersley and Broadcasting Pioneers)
BT Archives (PO Cable Ship, BT Tower, System X Eqpt)
BT Educational Dept (Telephone Pioneers)
Science Museum, London (Telegraph Cable Map, Microchip)
Northern Telecom (Travelling Wave Tube)
Motorola (LS Integrated Circuit)
Muirhead Systems (Facsimile Eqpt)
ATT Co., USA (Various illustrations from Bell Laboratories'
Record and the *History of Science and Engineering in the Bell System*).

The author's thanks are also due to Mr Iain Vallance, former chairman of the British Telecom Board, and Dr Alan Rudge, former BT managing director, for making available the BT Technical Library, Martlesham, BT Archives London, and the services of the BT Design Centre, Martlesham in preparing illustrations.

Finally, the present publication is a revised and updated version of *The Communications Miracle*, published by Plenum of New York in 1995. The new material presented includes Chapter 20 'The History and Growth of the Internet' and Chapter 21 'The Development of the Mobile Radio Service'. In updating references, use has occasionally been made of sources in the World Wide Web, where these are particularly relevant.

Chapter 1

Introduction

1.1 Today's world of telecommunications and broadcasting

The present-day world telecommunication network is the most complex, extensive and costly of mankind's technological creations and, it could well be claimed, the most useful. Together with sound and television broadcasting, telecommunications provides the nervous system essential for the social, economic and political development of mankind. It enables any user of the 1000 million or more world total of telephones in homes and offices, ships, aircraft and motor cars to communicate with any other, regardless of distance.

It provides the means for the fast distribution of documents and 'electronic' mail; it enables man to communicate by data transmission over the network with remote computers and databanks, and computer with computer. The long-distance transmission of television signals is now a commonplace, even from far distant planets of the solar system. The use of television for business conferences between offices in distant locations is increasing as more wide-bandwidth channels become available in the world telecommunication network.

In addition, a whole new art of information access, exchange and processing, known as information technology (IT) has developed whereby users in both home and office can gain access to virtually unlimited pages of information, visually displayed, from local or remote databanks via the telecommunication network, and interact with the displayed information or send messages on a person to person basis.

The future impact of IT linked to telecommunications could well be immense (1) by removing the need to travel to communicate, the vast waste of human and material resources needed to provide ever expanding rail and road facilities for countless millions of commuters every day from homes to city offices could be minimised; and (2) by diverting much office work to villages and small towns the quality of life could be enhanced and the rural economy sustained.

This vast development in the scale and range of telecommunication services and facilities has become possible mainly by a remarkable and continuous evolution in telecommunication system concepts and supporting technology, much of which has taken place in the last two or three decades, but of which the mathematical, scientific and conceptual origins can be traced back a century or more. As to the advances that telecomunications technology has made, one has only to compare the primitive slow-speed telegraphs, the pole-mounted copper wires and the manual telephone exchanges of the early 1900s (then the only means of intercity communication) with the advanced coaxial cable, microwave radio-relay, optical fibre and Earth-orbiting satellite transmission systems, and the fast computer-controlled electronic exchange switching systems handling voice, data and vision, in use today.

Device technology has made massive leaps forward from the thermionic valves of the early 1900s to the revolutionary invention of the solid-state transistor and microtechnology that puts millions of transistors and related devices on a single match-head-size microchip.

These advances in technology have not only made possible a vast growth in the scale and variety of telecommunication services, but have also enabled the cost to users to be progressively reduced in real terms.

As the distances over which signals need to be transmitted have increased, so new techniques such as pulsecode modulation and digital transmission have evolved, which greatly enhanced quality, flexibility and reliability.

The technology of broadcasting has also evolved from the mono-aural amplitude-modulation, long- and medium-wave sound broadcasting services introduced in the early 1900s to the stereophonic high-quality frequency-modulation VHF sound broadcasting and high-definition colour television UHF services, with Teletext (visually displayed alpha-numeric information), in the later 1900s. By the end of 2000 digital modulation had enabled many new television channels to be provided. Television broadcasting direct from satellites to the home is now commonplace.

These developments have not come about fortuitously; rather they have been initiated by the highly creative and original thinking of a number of individuals over a century or more.

It is the purpose of this book to seek to identify at least some of the scientists, mathematicians and engineers whose contributions can each be seen, in retrospect, to have provided an essential key or stimulus to the major developments that followed in later years, and to assess the significance of their individual contributions. However, while some of these innovators may have been inspired by visions of the future, it is certain that very few if any can have foreseen the immense impact that their work has had, and will continue to have, on telecommunications in the 21st century and beyond.

In selecting those on whose work attention has been focused, it has to be recognised that one may be doing less than justice to those who were providing the background from which those creative ideas grew, and to those who pursued those ideas to practicality and commercial success – and of course, judgement as to what is a major and specially significant contribution may itself be highly subjective.

Nevertheless, looking back over the long history of telecommunications development, it would seem that there are real peaks of creative achievement as clearly recognisable as Darwin's in the theory of evolution and Newton's in mathematical physics and astronomy. It is as a tribute to these men, and in the hope that it may, in some degree, inspire a new generation of 'telecommunicators' to further achievement, that this book has been written.

1.2 Innovation, invention and communication

As Edward de Bono has observed, 'There is nothing more important than an idea in the mind of man – man's achievements are based on man's ideas'. [1]

The history of technology has demonstrated time and again that it is the innovative and creative idea in the mind of an individual – sometimes stimulated by others working in the same field but often working alone – that has provided the start of developments that can now be seen to have been of major importance in the onward march of civilisation.

Most innovations and inventions in telecommunications and broadcasting have originated from a combination of need, e.g. to communicate person to person at a distance, or to a mass audience, and of recognition of the relevant scientific or engineering principles and possibilities, and at times chance observation of phenomena.

Inventions, and inventors, have not always been welcomed or understood by those in authority. For example, the Italian government showed little interest in Marconi's work on wireless, the British Post Office initially failed to recognise the value of Strowger's invention of the automatic telephone exchange, and even the transistor was regarded by one leading scientist as 'unlikely to replace the thermionic valve'.

It may be that, as the scale of telecommunications and broadcasting has expanded to cover the globe, the scope for individual contributions to techno-logical development has been overlaid by teamwork and the massive power of the large corporations; nevertheless, the role played by individuals remains cru-cial and deserves to be recognised. One message from this survey of innovation and invention stands out with clarity: no worthwhile invention was ever created because a team of cost study experts and accountants said 'you must invent this or that'.

1.3 The changing scene

With the invention of the telephone in 1876 many operating companies sought to exploit the telephone in and near large cities. However, it soon became appar-ent that this was not the way to run a national telephone network – some companies refused to connect their customers with those of rival companies. In addition, problems arose in providing economically for long-distance, i.e. intercity, connections. In the UK this chaotic situation was resolved in the early 1900s by the creation of a national telephone company under the aegis of the British Post Office (BPO). This pattern of 'Post, Telephone and Telegraph' (PTT) became the norm in most European and some other countries throughout most of the 20th century. The technical standards and operating procedures of the various national networks were co-ordinated by the International Telecommunication Union, Geneva, an organ of the United Nations.

The need, in the interests of efficiency, to run the BPO commercially, rather than as a government department, was recognised by the Post Office Act of 1969 which created separate businesses for Post and Telecommunications, each with its own board of control.

This process was taken further by the British Telecomunications Act of 1981 which established BT as a public corporation and introduced a degree of competition into the provision of telecommunication services. Relations between BT, competing companies and the interests of customers were regulated by the government-appointed body Oftel, the 'Office of Telecomunication Regulation'. [2] [3]

By 2000 the introduction of competition in the UK had begun to involve 'unbundling of the local loop', i.e. enabling competing services to be provided via BT's copper pairs between customer premises and BT local telephone exchanges.

In the USA the ATT Co.mpany (ATT Co.) had, during most of the 20th century, provided that country with the most extensive and efficient telephone system in the world. It owed its strength to the combination of Bell Laboratories (research and development), Western Electric (equipment design and manufacture) and the regional Bell Operating Companies (customer service).

ATT/Bell has a proud record of innovation in technology and system concepts, including the transistor, electronic exchanges, microwave radio-relay systems, submarine telephone cables and satellite communications.

However, in 1982 anti-trust legislation in the USA began a massive dismembering of ATT, initially of the regional Bell operating companies. In 1996 ATT itself divided into three companies, one primarily concerned with its traditional field of long-distance communications, in order to compete more effectively in a deregulated telecommunications industry.

To the author it would appear that, unless carefully regulated, competition and privatisation in what are essentially public service utilities such as telecommunications and transport – notably railways – are not always to the benefit of the customer. (The UK is still suffering from over-privatisation of the railways.) They are in a different category from competition in the manufacture and sale of consumer goods where errors tend to be short-lived and self-correcting.

The profit drive in the provision of competing telecommunication services may well lead, by unnecessary multiplication, to uneconomic use of national resources, confusion to and dissatisfaction of, the users. Furthermore, the linkage to advertising in some services, such as television and video, degrades the programme quality and viewer satisfaction with the service offered. A true 'view on demand' television/recorded video service, free from advertising and provided via electronic video libraries, is an ideal yet to be realised (see Chapter 19).

References

1 *Eureka: an Illustrated History of Invention*, ed. Edward de Bono (Thames and Hudson Ltd, Cambridge, 1974).
2 *Events in Telecommunication History*, BT Archives (www.bt.com/archives/history).
3 OFTEL, 'Office of the Telecommunication Regulator' (www.oftel.gov.uk).

Chapter 2

Creators of the mathematical and scientific foundations

Telecommunications depends heavily on scientific discoveries and advances in mathematics made by a small number of scientists and mathematicians in Europe during the mid-19th century, among whose names those of Volta, Oersted, Ampère and Ohm, Faraday, Maxwell, Hertz and Heaviside are outstanding.

In much of their work these men were exploring hitherto uncharted areas of knowledge and grappling with difficult concepts as to the nature of electricity and magnetism. They sought and found means for defining, and expressing in quantitative terms, the various electrical and magnetic phenomena they encountered, and evolved new terms such as voltage, current, impedance, lines of force, field strength and electromagnetic waves. By so doing, the prediction of performance of electromagnetic systems became possible and the bases for electrical design were laid.

The lives and work of these men have been described in detail and in depth in Rollo Appleyard's *Pioneers of Electrical Communication*, published in 1930 but now out of print. Much of the account in this chapter of the early pioneer scientists and mathematicians is based on Appleyard's book. Use has also been made of the summaries of the work of the pioneers prepared by the British Telecom Education Service. [1] [2]

Alessandro Volta (1745–1827)

Alessandro Volta has a lasting claim to fame in that the electrical unit of potential, the 'volt', has been named after him.

Volta was born in 1745, the son of a well-to-do merchant of Venice. His education took place mainly at the Royal Seminary in Como, where he had leanings towards poetry. However, a strong interest in chemistry and electricity took over and by 1779 he had been elected to the Chair of Physics at Padua University.

He travelled extensively, visiting England, France and Germany – this was

Volta (1745–1827)

Ohm (1789–1854)

Ampère (1775–1836)

Figure 2.1   Portraits of Volta, Ohm and Ampère

remarkable since the Napoleonic wars were then in progress and Europe was in a turmoil. Many of his contacts in these travels were with philosophers and leading men of science, e.g. Voltaire, Priestly, Lavoisier and Laplace. He was also acquainted with Luigi Galvani of Bologna and knew of Galvani's discovery that a frog's muscles twitched when joined by a pair of dissimilar metals. From this chance observation grew Volta's most significant invention the 'Voltaic Pile' – a series connection of pairs of dissimilar metals separated by membranes moistened with salt water that constituted an electric battery.

This invention he described in his paper 'On the Electricity Excited by the Mere Contact of Conducting Substances of Different Kinds', recorded in the *Philosophical Transactions* of the Royal Society, London in 1800. Although Volta produced a number of other electrical inventions, e.g. the 'electrophorous', which generated static electricity by friction on an insulator, and a 'condensing electroscope' for detecting small charges of static electricity, there is little doubt that the 'voltaic pile', the forerunner of the modern multicell battery, was by far the most valuable.

André Marie Ampère (1775–1836)

The name of this distinguished French scientist has been honoured for all time by designating the unit of electrical current the 'ampere'.

Ampère was born in 1775 when France was moving towards the Revolution that began in 1789. His father died on the scaffold and Ampère's early life was much clouded by this traumatic event.

However, he sought solace and became interested in mathematics and studied at the École Centrale in Bourg, where he set up a laboratory for the study of the physical sciences. In 1804 he was awarded a professorship at the École Polytechnique in Paris. His scientific work covered a broad front and he made many original contributions in mathematics, mechanics, electricity, magnetism, optics and chemistry. His most significant contribution in electricity and magnetism was concerned with the mechanical forces between current-carrying conductors, and the formulation of the laws governing such forces, described in a presentation to the Académie des Sciences in 1820. In so doing he followed on from Oersted's observations on the deflection of a magnetic needle by a current-carrying conductor.

A useful byproduct of Ampère's work in electromagnetism was his 'astatic galvanometer', in which the deflecting coils were so arranged as to neutralise the effects of external magnetic fields such as the Earth's, thereby increasing the useable sensitivity.

Ampère's formulation of the laws relating to the forces generated between current-carrying conductors later proved to be of major importance in the design of telegraph apparatus, telephone receivers and loudspeakers.

Georg Simon Ohm (1789–1854)

The name of G.S. Ohm, like those of Volta and Ampère, has by international agreement been used to designate an important electrical unit, in his case that of resistance, the 'ohm'. To quote from Ohm's biographer Rollo Appleyard:

A century ago (early 1830s), the science and practice of electrical measurement and the principles of design for electrical instruments hardly existed. With a few exceptions, ill-defined expressions relating to quantity and intensity, combined with immature ideas of conductivity and derived circuits, retarded the progress of quantitative electrical investigations. Yet, amidst this confusion, a discovery had been made that was destined to convert order out of chaos, to convert electrical measurement into the most precise of all physical operations, and to aid almost every other branch of quantitative research. This discovery resulted from the arduous labours of Georg Simon Ohm.

G.S. Ohm, who was to achieve fame as a physicist, was born in 1789 in the university town of Erlangen, in Bavaria; a brother named Martin Ohm was born in 1792 and became a noted mathematician. Georg Ohm studied mathematics and physics at Erlangen University; however, he had also, for reasons of economy, to teach, which at that time he found irksome.

In the early 1800s conditions in Bavaria became difficult as the struggle against Napoleon developed. Ohm left his native Bavaria in 1817 to settle in Cologne, where he obtained a Readership at the university, and found both freedom and appreciation. He had developed a growing interest in studying the conductivity of metals and the behaviour of electrical circuits and, to pursue these studies more effectively, he gave up teaching in Cologne and went to live in his brother's home in Berlin. Another objective was to prepare a thesis that could help towards achieving a university professorship.

After intensive studies he published *Die galvanischekette, mathematisch bearbeitet*, which established for the first time the relationship between voltage (potential), current and resistance in an electrical circuit, immortalised in the simple relationship

$$I = E/R.$$

Of considerable importance was the concept due to Ohm that the effective potential in a circuit was the sum of the individual potentials, and the effective resistance was the sum of the resistances of the component parts of the circuit.

There was for a time opposition to Ohm's Law by one or two physicists, and by the philosopher Hegel who disputed the validity of Ohm's experimental approach. Eventually Ohm's triumph came in 1841 when the Royal Society of London awarded him the Copley Medal for his work.

Recognition followed in Bavaria and Germany, by numerous professional and academic appointments, including one as adviser to the State on the development of telegraphy.

The applicability of Ohm's Law to electrolytes and thermoelectric junctions, as well as to metallic conductors, was in time recognised, and it remains today the most widely used and valued of all the laws governing the behaviour of electrical circuits.

Hans Christian Oersted (1777–1851)

Oersted's major contribution to electrical science was the demonstration from the simple observation of the deflection of a magnetic compass needle that a magnetic field existed around a current-carrying conductor. This experimental proof of the close connection between electricity and magnetism stimulated in-depth studies by many other scientists, including Faraday, Clerk Maxwell and

 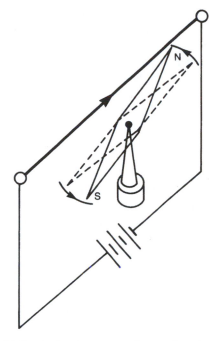

(a) Hans Christian Oersted (1777–1851)   (b) Oersted's compass needle experiment

Figure 2.2   Oersted and compass needle

Heaviside; in recognition of his work the unit of magnetic field strength was named the 'oersted'.

Oersted was born in 1777 at Rudkjobing, a small town on the Danish island of Langeland, where his father was an apothecary. Educational facilities on the island were limited; Hans and his young brother Anders received their primary education largely through the diligence and encouragement of their parents.

Through their hard work they gained places at Copenhagen University, where Hans was awarded a PhD in 1797. His interests were initially in literature and philosophy, and included philosophical studies seeking a unity between natural phenomena such as light, heat, chemical action, electricity and magnetism.

Volta's invention of the voltaic pile, which provided for the first time a continuous source of electrical current, inspired Oersted to concentrate on physics, and in 1806 he was appointed Professor of Physics at the University of Copenhagen. He travelled extensively in Germany and France, discussing experimental philosophy with men of science and forming long friendships. One such friendship was with Johann Wilhem Ritter (1776–1810) who had invented a battery (or secondary cell) which could be recharged time and again from a voltaic pile.

The claim that a magnetic compass needle could be deflected when a current-carrying conductor was in its vicinity was first reported in 1820, following an almost chance observation in the course of a classroom lecture. What is important, however, is that Oersted's earlier philosophical studies on the essential unity of natural phenomena such as electricity and magnetism made him realise immediately the profound significance of this seemingly chance observation.

Oersted wrote an account in Latin (then the normal language for communicating scientific discoveries) of theories of his compass needle experiment and his conviction that every voltaic circuit carries with it a magnetic field, and sent copies to societies and academies in all the capitals of Europe. He received many honours in recognition of his work, including like Ohm, the Copley Medal of the Royal Society in London.

While Oersted's discovery had an early application to the electric telegraph, as for example Wheatstone's needle telegraph, perhaps the most valuable outcome was the stimulus it gave to the mid-19th-century scientists, physicists and mathematicians who were seeking a unified electromagnetic theory.

Michael Faraday (1791–1867)

Michael Faraday's discoveries in electromagnetism provided the scientific basis for devices that were to become the essential keys to a wide range of new industries, ranging from induction coils and transformers to motors and dynamos essential for the generation, distribution and utilisation of electrical power, as well as for telecommunications. His name has been commemorated in the electrical unit of capacitance, the 'farad'. [3]

Michael Faraday was born in London in 1791. His early education was in a common day school, amounting to little more than the elements of reading, writing and arithmetic. At the age of 13 he became an apprentice to a bookseller, which enabled him to learn bookbinding. However, more importantly, it gave him an opportunity to read extensively, reading which included the articles on electricity in the *Encyclopaedia Britannica*.

When he was 21 he attended Sir Humphrey Davy's lectures on heat, light and the relationship between electricity and chemical action, at the Royal Institution in London. He became Davy's assistant and learned much from him, including the preparation of his own scientific papers and the art of public lecturing.

Faraday's interest in electromagnetism was stimulated by news of Oersted's compass needle experiment, which revealed that there was a magnetic field around a current-carrying conductor which could deflect a magnetic needle. From this Faraday deduced the principle of, and demonstrated in 1821, a simple electric motor in which a wire carrying a current rotated around a fixed electromagnet.

He was elected a Fellow of the Royal Institution in 1824, but not without some opposition from his mentor Davy. By 1826 he again turned his attention to electromagnetism, studying the work of Oersted, Ampère and Ohm. He reasoned that:

If an electric current in a conductor produces a magnetic condition, why should not a conductor near a magnet show an electric condition?

Michael Faraday (1791–1867)

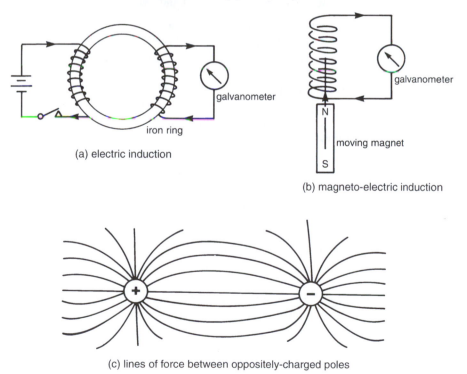

(a) electric induction

(b) magneto-electric induction

(c) lines of force between oppositely-charged poles

Figure 2.3   Faraday and experiments

This line of thought led Faraday to a series of experiments in 1831 that were later to have remarkable results. He gave much thought to what was happening in the field around magnets and current-carrying conductors, visualising 'lines of force' that linked opposite magnetic poles, or electric charges, and which exercised a pull or tension along such lines. This concept later enabled mathematical physicists such as Clerk Maxwell to give quantitative expression to electromagnetic phenomena.

From his experiments Faraday deduced the fundamental law of electro-magnetism:

Whenever a closed conductor moves near a magnet in such a manner as to cut across the magnetic lines, an electric current flows in the conductor.

His experiments led Faraday to describe two groups of electromagnetic phenomena:

(1) Those in which a varying current in one coil produces transient currents in a second coil.

This he called 'electric induction'. It was the basis of induction coils used later, for example by Edison in the telephone, and the alternating-current transformers used in the telecommunications and electrical power industries.

(2) Those involving relative motion between a magnet and a conductor.

This he called 'magneto-electric induction'. It was the basis of, for example, the telephone receiver in telecommunications and the loudspeaker in broadcasting, and it gave birth to electric motors and dynamos for the electrical power industry.

Faraday later turned his attention to the decomposition of liquids by electricity, creating new terms such as 'electrolyte' and 'electrolysis', 'anode' and 'cathode' – the latter were subsequently to have wide currency in the thermionic valves used in telecommunications.

In 1855, when he was 64 years old, Faraday completed his sizeable work *Experimental Researches*, embodying the experiments, theories and results derived from 24 years of intensive research and study.

In 1867 Faraday died peacefully in his study in a 'grace and favour' house at Hampton Court given to him by Queen Victoria in recognition of his scientific achievements. He was a remarkably modest man, refusing many honours and financial rewards for his work. Among his many friends were Wheatstone and Cooke, who later achieved fame for their contributions to the invention of the electric telegraph.

His happy marriage and profound religious convictions gave him a serenity of mind that must have helped him greatly in his scientific work. However, he is said not to have thought of himself as a 'scientist', but rather a 'philosopher', i.e. a student of the phenomena of nature.

Oliver Heaviside (1850–1925)

Oliver Heaviside's contributions to electromagnetic theory are now recognised as major advances in understanding the principles of operation and design of long telegraph and telephone cables, but in their day they were considered as most abstruse and beyond the understanding of ordinary mortals. His work provided, for the first time, an accurate basis for determining the maximum speed of signalling on telegraph cables and the conditions for achieving a 'distortionless' cable; his studies of the transmission of telephony on long cables revealed the possibility of improved speech quality by 'loading' the cable, i.e. adding inductance at intervals. In addition, his work on electromagnetic wave propagation revealed the possibility of round-the-world transmission by radio waves guided between the Earth or sea and an electrified layer in the upper atmosphere, now known as the 'Heaviside Layer'.

To achieve his objectives Heaviside had to invent highly unconventional mathematical methods and symbolism, including the use of 'operators', much of which he arrived at intuitively and to which the mathematical purists objected for alleged lack of rigour. Even his contemporaries working on electromagnetism found his work difficult to understand; the editors of the learned society journals to whom he looked to publish his work found him very awkward to deal with as a result of his unwillingness to adapt his manuscripts to the needs of their readers.

During a half-century of intensive theoretical work Heaviside wrote a great deal. Much of it is published in his *Electrical Papers* (2 vols) and *Electromagnetic Theory* (3 vols); other material has been preserved in the manuscript collection held at the Institution of Electrical Engineers, London, together with letters to his contemporaries, including Clerk Maxwell, Faraday, Lord Rayleigh and Professor G.F. Fitzgerald (Dublin).

Fitzgerald wrote in a review of Heaviside's Electrical Papers:

They teach a sound theory of telegraphs and telephones . . . which there is every prospect may lead to vast improvements in telegraphy and may even make it possible to work a telephone across the Atlantic.

Heaviside's theoretical prediction of the value of inductive loading on long cables was further developed and demonstrated practically by Michael Idvorsky Pupin (1858–1933). Pupin was born in what was Yugoslavia in 1858; he arrived in the USA as a penniless immigrant and later graduated from Columbia University. After studying in Germany under Helmholtz and Kirchoff, Pupin returned to the USA and became Professor of Electro-mechanics at Columbia University.

It is interesting to note that, prior to the outbreak of the Second World War in 1939 and before deep-sea repeaters were practicable, the Bell Telephone Laboratories in the USA had designed an inductively loaded and repeaterless transatlantic cable providing a single telephone circuit; implementation of the project was postponed by the war, and replaced in 1956 by a cable with valve-operated repeater/amplifiers providing 36 telephone circuits (see Chapter 10).

Figure 2.4   Heaviside and Pupin

Heaviside was born in Camden, London in 1850. Details of his early educa-
tion are lacking. After leaving school he joined the Great Northern Telegraph
Company, perhaps through the interest of his uncle Sir Charles Wheatstone, and
was also employed by the Anglo-Danish Cable Company. An elder brother,
Arthur West Heaviside, was a divisional engineer with the British General Post
Office.

Oliver and Arthur shared an interest in telegraphy and together carried out a
number of experiments. A memorandum of 1873 records how they demonstrated 'duplex telegraphy', i.e. simultaneous bothway transmission, over an
artificial line with a 'needle' telegraph. Later they demonstrated duplex telegraphy over an actual line between Newcastle and Sunderland. Their experiments
at this time included work on microphones, which perhaps later stimulated
Oliver to include telephony in his theoretical studies.

In 1882 Oliver Heaviside began his famous series of papers in the *Electrician*
on the theory of propagation of electromagnetic waves on conductors and in
space. This work is notable, *inter alia*, for its establishment of the principle of
duality between electricity and magnetism. Hertz's discovery in 1886 of the wave
character of electromagnetic radiation stimulated Heaviside to further theoretical studies, and by 1888 he was contributing to the *Philosophical Magazine* on
this subject.

In 1889 Heaviside and his parents moved to Paignton in Devon, a move which
perhaps isolated him from experimental work. This, his parents' death and his
own increasing deafness, made him more and more a recluse, concentrating
increasingly on his theoretical studies. His working habits became more
strange, involving long hours, often into the early morning, spent in a closed,
smoke-filled room.

He was autocratic by nature, highly intolerant of poseurs and those who failed to understand his work. However, he had superb powers of concentration, a remarkable intuitive perception of the nature of electromagnetism and an ability to translate this into theoretical statements of immense value for quantitative design.

Heaviside died in relative poverty in Torquay, at the age of 75. He was a member of the Society of Telegraph Engineers, a forerunner of the present Institution of Electrical Engineers, London, of which he was made an honorary member. Fittingly, he was awarded the IEE's first Faraday Medal. To him, the science of telecommunications owes a massive debt, unrewarded in his lifetime.

Heinrich Rudolf Hertz (1857–1894)

It can be said that the work of the earliest pioneers of electromagnetic studies culminated in 1886 in Karlsruhe when Heinrich Hertz demonstrated experimentally the wave character of electrical transmission in space, as well as along wires. However, Hertz himself was the first to acknowledge that his work was significant because it verified the theoretical predictions made by others, including Maxwell, Faraday and von Helmholtz. In a sense it is fortunate as well as fortuitous that Hertz's experimental work, because of the convenient dimensions of his laboratory apparatus, involved centimetric wavelength radio waves. These could be readily reflected by sheets of metal, focused by parabolic mirrors and deflected by dielectric prisms. His work thus had application not only to the propagation of long- and medium-wavelength radio waves first exploited by Marconi, but also later prepared the way for microwave radio-relay systems, radar and satellite communications. In honour of Hertz's work, the unit of frequency, the cycle per second, has been internationally named the 'Hertz'.

Hertz was born in Hamburg in 1857, the eldest son of a Jewish advocate and a member of a prosperous family of middle-class merchants. His grandfather

Figure 2.5   Hertz and Maxwell

studied natural science as a hobby and left his grandson a small laboratory where the young Hertz carried out simple experiments in physics and chemistry.

He demonstrated an early ability in languages and became proficient in English, French and Italian, a skill which must have helped him to understand the work of his contemporaries in other countries. His most formative studies were at the University of Berlin under von Helmholtz and Kirchoff, where he carried out individual research on a largely philosophical question concerning the extra current manifested when an electric current starts or stops – a problem also studied by Maxwell. The university awarded him a doctorate for his thesis, with the added and unusual distinction of 'magna cum laude'.

In 1880 Hertz became a demonstrator in physics at the University of Berlin under the direction of von Helmholtz, leaving in 1883 to take up a professorial post in the University of Kiel, where he studied electromagnetism intensively, with particular attention to Maxwell's work in this area.

In 1885 he became Professor of Experimental Physics at the Technische Hochscule in Karlsruhe. It was here that in 1886 he noted that the discharge of a Leyden jar (a charged capacitor) through one of a pair of spiral coils caused a spark to pass across a short airgap between the ends of the other coil. Later experiments used a spark coil, similar to a trembler bell mechanism, to generate continuous sparks.

It had been pointed out by Oliver Heaviside, in the *Philosophical Magazine*, 1877, that the oscillatory nature of the discharge of a charged capacitor in association with a self-inductance was first disclosed by Joseph Henry in 1842. Sir William Thomson (later Lord Kelvin) had also studied the theory of this phenomenon, reported in the *Philosophical Magazine*, 1853, and experimental proof had been shown by von Bezold in 1870. Although Hertz was unaware of Bezold's work at the time of his first experiments, he was the first to acknowledge this when reporting his own discoveries. An important step forward was made when Hertz demonstrated the principle of resonance between the transmitting and receiving circuits – initially by dispensing with the Leyden jar in his spiral coil experiments – which, by making these circuits identical, increased the distance over which signals could be detected.

A further highly significant step was to realise that the detection distance was more favourable than the Newtonian law of inverse squares, in fact more nearly an inverse distance relationship.

His invention of the 'linear' oscillator – two straight metal rods terminated by metal spheres providing capacitance – was the forerunner of dipole aerials used in present-day radio, radar and television systems. With it he was able to demonstrate that the waves produced were 'linearly polarised', i.e. the oscillations of the electric component of the radiated field were parallel to the rods of the oscillator. A receiver with rods at right angles to the transmitting rods would pick up virtually no signal. This principle is valuable for the avoidance of interference between radio systems sharing the same frequency, thereby aiding economy in spectrum utilisation.

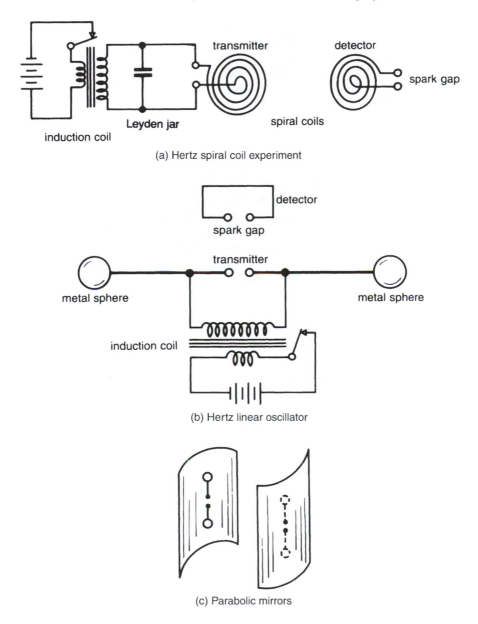

(a) Hertz spiral coil experiment

(b) Hertz linear oscillator

(c) Parabolic mirrors

Figure 2.6　Hertz's experiments

The demonstration that short-wavelength radio waves could be efficiently concentrated into beams by parabolic reflectors of dimensions comparable with, or greater than, the wavelength later led to directional aerials providing power gain, i.e.more effective transmission and longer range than is possible between dipole aerials by themselves. The radio frequencies used in Hertz's experiments were of the order of 1GHz, the wavelength being about 30 cm and the parabolic reflectors about 100 cm in diameter.

Hertz's discoveries and his studies of their relationship to Maxwell's theories were published in 1891 in a treatise written in German and dedicated to von Helmholtz; this was translated into English by Professor D.E. Jones, with a preface by Lord Kelvin, under the title *Electric Waves, being Researches on the Propagation of Waves of Electric Action With Finite Velocity Through Space*.

The reference to 'finite velocity' in this title is interesting since it is in conflict with the concept of instantaneous 'action at a distance' supported by von Helmholtz, but is in accord with Maxwell's theories of electromagnetic wave propagation. Hertz was honoured by the Royal Society, London, which presented him with the Rumford Medal. He is reported as 'being of an amiable disposition, genial, a good lecturer, possessed of singular modesty who – even when speaking of his own discoveries – never mentioned himself.'

Hertz died in 1894, at the age of 37. He can have had no premonition of the ultimate value of his work, for, when asked whether there was a possibility that the waves that came to be called 'Hertzian' could be used for telegraph and telephone communication between countries, he replied 'no, it would need a mirror as large as a continent'.

Hertz can, nevertheless, justly be regarded as the 'Father of Radio Communication'.

James Clerk Maxwell (1831–1879)

Maxwell's biographer Rollo Appleyard has said that 'Throughout the history of natural science there is no name except Newton's more honoured by men of science than that of James Clerk Maxwell'.

It was Maxwell's remarkable achievement to have demonstrated, by reasoning and pure mathematics, embodied in *Maxwell's Equations*, the fundamental relationships between electricity, magnetism and wave propagation that underly all radio and cable communication. He showed that light is, like radio waves, an electromagnetic phenomenon and both have a maximum velocity of 300 million metres per second. His work provided a theoretical basis for believing that radio waves, like light, could be focused by conducting reflectors as Hertz later demonstrated practically. *Maxwell's Equations* remain a fundamental starting point for much electrical design, not only for cables both electric and optical but also the microwave components and aerials used in present-day radio-relay, satellite communication and radar systems.

Maxwell also had an interest in colour vision, shown by the experiments he conducted to demonstrate the nature of light, and which later had a bearing on the development of colour television.

He was born in Edinburgh in 1831; his early education took place at Edinburgh Academy where his talent won him prizes in English and mathematics. At 16 he entered the University of Edinburgh where his mathematical abilities flourished under the tuition of Sir William Hamilton, inventor of the mathematical concept 'quaternions'. In 1850 he went to Cambridge University, where he

(a) James Clerk Maxwell (1831–1879) (US Library of Congress)

Maxwell's equations of the electromagnetic field, expressed in m.k.s. units are:

$$\text{curl } E = -\dot{B}, \text{ div } B = 0$$
$$\text{curl } H = \dot{D} + J, \text{ div } D = \rho$$

where

$$B = \mu H, D = \varepsilon E$$
$$\mu = \mu_0 K_m, \varepsilon = \varepsilon_0 K_e$$

and K_m and K_e are respectively the magnetic and electric specific inductive capacitances, and μ_0 and ε_0 the magnetic and electric inductive capacitances of a vacuum. J = current density and ρ = charge density.

(b) Maxwell's electromagnetic field equations

Figure 2.7   Maxwell and equations

came under the influence of Sir George Stokes, mathematician and physicist. In 1854 he gained a Fellowship at Trinity College and besides lecturing took an interest in Faraday's work on electromagnetism. He became for a time a Professor at King's College, London, where he developed a close friendship with Faraday.

In the period 1850–1865 there were many advances in submarine cable telegraphy, bringing both data and problems for Maxwell to solve. During a period of illness and temporary retirement to his native Scotland he began his great work *Electricity and Magnetism*, completed in 1873. In 1871 he was given the Chair of Experimental Physics at Cambridge University, a move which later involved him in the creation of the world-famous Cavendish Laboratory.

The quality of Maxwell's thinking, and his outstanding ability to crystallise subtle electromagnetic phenomena into meaningful quantitative terms can be exemplified by the following statement of two of his four theorems:

1. If a closed curve be drawn embracing an electric current, then the integral of the magnetic intensity taken round the closed curve is equal to the current multiplied by 4 pi.
2. If a conducting circuit embraces a number of lines of magnetic force, and if for any cause whatever the number of these lines is diminished, an electro-motive force will act round the circuit, the total amount of which will be equal to the decrement of the number of lines of magnetic force H in unit time.

Maxwell's theory owes something to Ohm's Law, and a great deal to Faraday who saw 'lines of force' where other mathematicians saw 'action at a distance'. When he applied his equations to electromagnetic wave propagation Maxwell came to the conclusion that:

The velocity is so nearly that of light that it seems we have strong reason to conclude that light itself including radiant heat, and other radiations if any is an electromagnetic disturbance in the form of waves propagated through the electromagnetic field according to electromagnetic laws.

It is reported that Maxwell was 'genial and patient, had great power of concentration, considerable knowledge and discrimination in literature, was a rapid reader and had a retentive memory – he loved his friends, his dog and his horse'.

He died in 1879, aged only 48. To those proceeding to extend his victories along the road of electrical communication he left his 'sword and chariot' – his equations and theories. He also left an example of individual thought and achievement and a plea for fellowship between all men of science.

William Thomson, Lord Kelvin (1824–1907)
William Thomson was knighted when 34 years old for his part in the laying of the first transatlantic telegraph cable (see Chapter 3). He was given a peerage by Queen Victoria in 1896, becoming Lord Kelvin, and was regarded in his day as the foremost physicist and electrical engineer in the world. [4]

During his long, happy and vastly productive life, William Thomson contributed notably to thermodynamics, electricity and magnetism, electrical engineering, telegraphy, hydrodynamics, navigation and mathematics.

He was born in Belfast, Northern Ireland in 1824. His father James Thomson was Professor of Mathematics at Glasgow University where his son William later became Professor of Natural Philosophy for almost the whole of his working life. William matriculated at the remarkably early age of ten, attending his father's lectures and showing considerable aptitude for languages, logic and mathematics. He entered Cambridge University (Peterhouse) at the age of 17, specialising in mathematics.

William took an active part in university life, being a good oarsman and a performer in the Cambridge Musical Society. Like many another student he had difficulty in making ends meet and his father – a good Scot – frequently wrote to adjure him to be more careful with his money!

After graduating from Cambridge and studying in the Regnault laboratories

Lord Kelvin, Professor of Natural Philosophy at the University of Glasgow, 1846–99; Chancellor of the University of Glasgow, 1904–7. From a portrait in the University of Glasgow by Sir Hubert von Herkomer, painted in 1891.

The 'cable' galvanometer used to receive the first transatlantic signals in 1858. From the museum collection of the Natural Philosophy Department at the University of Glasgow.

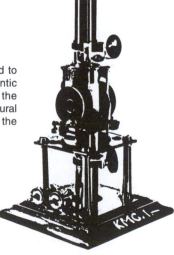

Figure 2.8   Kelvin and cable galvanometer

in Paris for a time, William was awarded the Chair of Natural Philosophy at Glasgow University at the age of 22, where he remained as Professor of Physics for 53 years. In spite of this long tenure of this post his versatility and creativity remained at an astonishingly high level throughout.

Very early in his career Thomson became interested in the problems of heat flow, and his first paper on this was published when he was 15 in the *Proceedings of the Cambridge Mathematical Society*. His work on heat as a form of energy was both creative and profound, and led to the establishment of an absolute zero of temperature at minus 273 degrees Centigrade as a datum for the scale of temperature later designated, in his honour, as 'Kelvin'. He also contributed to the formulation of the keystone *First and Second Laws* of thermodynamics.

The First Law demonstrated that, taking into account gains and losses of energy, including heat, in a closed energy system there was at all times a balance of energy in the system.

The *Second Law* proved that heat can only change into useful work in flowing from a higher temperature T1 to a lower temperature T2, and not in the reverse direction (sometimes called 'Time's Arrow', which has significant philosophical implications). It also enabled a formula $(T1 - T2)/T1$ to be derived for the optimum efficiency of heat engines that set a target for improving the often grossly inefficient steam and other engines.

Thomson's interest in the relation between heat and electricity led him to study the 'Seebeck effect', which relates to the generation of electricity by heat applied at a junction between a pair of dissimilar conductors, and the 'Peltier effect' that is concerned with the cooling at such a junction when carrying an electric current. His work showed the quantitative relationship between these two effects and revealed a third effect *Thomson heating*. These results were to prove of major importance in the design of refrigeration systems.

His electrical inventions were numerous and valuable – they included an 'electrostatic voltmeter', a 'quadrant electrometer' and a 'balance for the measurement of electric currents'. But perhaps the most useful from a telecommunication point of view was the highly-sensitive 'mirror galvanometer' that played such an important role in the first trans-Atlantic telegraph cable, see Chapter 3.

He also invented means for correcting the magnetic compass on board ships from errors due to the ship's magnetic field and improved the mounting of the compass to minimise the effect of rolling, further assisting in overcoming the difficulties of submarine cable laying.

Sir J.J. Thomson (no relation of William Thomson), then a Professor at the Cavendish Laboratory, Cambridge, said of Kelvin:

Modern wireless telegraphy, telephony and broadcasting depend on a result published by Kelvin in 1853.

The result referred to appears in a paper by William Thomson on 'Transient Electric Currents' in the *Philosophical Magazine*, June 1853 in which he derived solutions for the behaviour of current in a damped oscillatory circuit, including the result that, when the damping is light, the resonant frequency is given by:

$$f = 1/2\pi\sqrt{LC}$$

where L and C are the inductance and capacitance of the oscillatory circuit. This equation is written in the mind, if not on the heart, of all good radio engineers today!

Kelvin's thinking was not however limited to the specific scientific problems of everyday life; his *Second Law of Thermo-dynamics*, with its implication of 'Time's Arrow', led him to speculate on the age of the Earth and indeed the Universe itself. Although his estimate of the age of the Earth was in error by a substantial margin, he nevertheless came to the conclusion that all things had a beginning (supported by the present-day 'big bang' theory) and that 'if a probable solution, consistent with ordinary course of nature, can be found, we must not invoke an abnormal act of Creative Power'.

Kelvin died in 1907, with many honours including Fellowships of the Royal Societies of Edinburgh and London (granted when he was only 22 and 26 years of age respectively). He was buried in Westminster Abbey, fittingly near the grave of Sir Isaac Newton.

Kelvin had a remarkable simplicity of character and a full measure of scientific modesty. At his Professorial Jubilee in Glasgow University in 1896, after he had played such a great role in the development of telegraphy and electrical engineering, he said rather wistfully:

I know no more of electric and magnetic force ... than I knew and tried to teach my students of Natural Philosophy fifty years ago.

2.1 The electron is discovered and the electronic age begins

Once the laws of electro-magnetism and the identity of light with electro-magnetic waves had been established, the next major advance was the discovery of the *electron*, a discovery that led to an understanding of the structure of atoms and the development of quantum physics, vital to the creation of new electronic devices such as the thermionic valve, the transistor and the microchip, the cathode-ray picture tube, lasers as sources of coherent light and light-sensitive photo-diodes.

Sir J.J. Thomson (1856–1940)
The first step – the discovery of the electron – was made by J.J.Thomson, the son of a Manchester bookseller, who entered a local college at the age of 14 intending to take up a career in engineering. However, he found experimental physics more interesting and in 1876 won a scholarship to Cambridge University where he remained for the rest of his life. In addition to his outstanding abilities in mathematics and physics he was a skilled administrator and his advance at Cambridge was spectacular. At the age of 27 he succeeded the distinguished physicist John Rayleigh as Professor of Physics and became responsible for the Cavendish Laboratory, which acquired world-wide fame for research in atomic physics under his direction. [5]

Sir J. J. Thomson (1856–1940)

Above: a discharge tube designed by Thomson, in which cathode rays were allowed to travel past electrically charged plates. The deflection of the cathode rays by electric and magnetic fields convinced Thomson that these rays consisted of particles, which we now know as electrons.

Sir Ernest Rutherford (1871–1931)

Rutherford's first notes on the structure of the atom.

Figure 2.9   Pioneers of the electronic age: Sir J. J. Thomson and Sir E. Rutherford

Thomson's discovery of the electron arose from his interest in cathode rays, the radiation from the negative electrode or cathode when an electric current passed between two electrodes in an evacuated glass tube. By measuring the deflections of the cathode rays in the presence of electric and magnetic fields of known strengths, Thomson was able to show that the cathode rays were streams of negatively charged particles he named 'corpuscles', later called 'electrons'. From the deflection of the cathode rays by electric and magnetic fields he deduced the ratio of the electric charge to the mass of the individual electrons, and showed that these were much smaller in size than the atoms themselves.

Support for Thomson's discovery came a few years later in 1886 when the German chemist Eugen Goldstein showed that the electron beam in a cathode-ray tube was accompanied by a stream of positive charges, which he called 'canal rays', moving in the opposite direction. These were 'ions', loosely-bound atoms stripped of their electrons. He further demonstrated that, if the cathode had a hole, the canal rays produced a luminous spot on a screen behind the cathode.

Although the modern cathode-ray tube uses an electron stream rather than positively-charged particles to produce waveform and picture displays, its origin can clearly be seen in the ideas initiated by Thomson and Goldstein in the late 19th century.

Thomson speculated about the structure of the atom in terms of charged particles, the negative ones 'electrons' being akin to his 'corpuscles'. This theme was later followed up by Rutherford, see below.

Thomson was one of the great intellectual celebrities of British science. He was also an inspired teacher and he gathered round him at the Cavendish Laboratory some of the most brilliant young minds of the day, many of whom became professors in the world's leading universities and seven won Nobel prizes. One of Thomson's former students was Sir Ernest Rutherford who later achieved much fame in the field of atomic physics.

Sir Ernest Rutherford (1871–1937)

Ernest Rutherford was born in New Zealand, where his father was a farmer. Ernest in his younger days shared the hard work of running the farm. He showed great promise as a student and won a scholarship at Canterbury College in Christchurch, New Zealand. He graduated with honours in 1882 and won another scholarship which took him to Cambridge University, where he met Sir J.J.Thomson who encouraged him to work on X-rays that had just been discovered by the German scientist Wilhelm Rontgen. For Rutherford this was the beginning of a lifetime interest in radioactivity and atomic structure. The former led to many studies of the properties of the emissions, i.e. alpha, beta and gamma rays, from radioactive elements in the uranium and thorium series of atoms which terminate in stable lead atoms. He expressed the exponential decay of the rate of emission from an active atom in terms of the 'half life', at which the rate of emission has been reduced to half.

In a classic series of experiments in 1909 he analysed the pattern of particles coming from a gold target bombarded by a beam of alpha particles, giving him an insight into the structure of atoms. He formulated a model in which the mass of the atom was concentrated in a positively-charged nucleus, around which electrons moved much as do planets around the Sun. This model was refined, notably by Niels Bohr, and provided a foundation for the quantum theory, [6] which enabled a wide range of apparently unconnected physical phenomena to be explained.

Rutherford's work shows the remarkable power of an original mind to find commonsense answers to complex scientific problems. His was a warm and humorous nature that won him many friends. His strong opposition to the rise of the Nazi party in Germany during the pre-War years led him to help many Jewish refugees to flee from the Nazi persecution. In 1931 he was honoured with the title of Baron Rutherford of Nelson. With J.J. Thomson he helped to create a vital foundation for today's electronic industries.

Lord Rayleigh (1842–1919)

The scientific career of John William Strutt, third Baron Rayleigh, covered a period when physical science was undergoing remarkable development at the hands of his contemporaries such as James Clerk Maxwell, Oliver Heaviside and Lord Kelvin. A man of vast erudition, broad interests and tremendous energy, he made important contributions to every branch of physics known in his day, including heat, light and sound, electricity, magnetism, and the properties of matter. [7]

Rayleigh's *Scientific Papers* (Dover Publications, New York, 1964) comprises over 400 writings. Perhaps the more important for their later impact on the development of telecommunications and broadcasting were the following:

Rayleigh's *Theory of Sound*, one of the great classics in the literature of physics, his work on *The Scattering of Light by Small Particles*, which laid a foundation for later theories of radio-wave propagation by scattering in the troposphere and ionosphere, and the determination of electrical standards, notably the ohm. His treatise *The Passage of Electric Waves Through Tubes, or the Vibrations of Dielectric Cylinders*, became a starting point for the theory of long-distance waveguide and optical-fibre transmission systems (see Chapters 16 and 17).

Lord Rayleigh was born John William Strutt, eldest son of the second Baron Rayleigh of Terling Place, Witham, in the county of Essex, England. From his earliest schooling days he showed an aptitude for mathematics and in 1861, when nearly 20, he entered Trinity College, Cambridge. Here his mathematical bent was stimulated by a famous mathematical 'coach' E.J. Routh, and Sir George Stokes, the Lucassian Professor of Mathematics who was also greatly interested in experimental physics. In the Cambridge Mathematical Tripos of 1865 Rayleigh emerged as Senior Wrangler and in 1866 was elected a Fellow of Trinity College.

The establishment of the Cavendish Laboratory at Cambridge with James

Clerk Maxwell as the first Professor of Experimental Physics provided a further stimulating background to Rayleigh's work. On Maxwell's untimely death in 1879 Rayleigh succeeded to the Professorship and initiated a comprehensive teaching programme in heat, electricity and magnetism, optics, acoustics, and the properties of matter – a pioneering effort that had a major impact on physics teaching in the UK and Europe.

However, much of his experimental work was carried out in his rather primitively equipped laboratory at his home in Terling Place, Essex.

He became President of the Mathematics and Physics Section of the British Association and Secretary, later President, of the Royal Society, London. In 1867 he was appointed Professor of Natural Philosophy at the Royal Institution of Great Britain, where Faraday had made his epoch-making investigations into electro-magnetism.

It was at the Royal Institution that Rayleigh began the famous 'Friday evening' discourses and the Christmas 'Children's Lectures' that are so popular today as a means for promoting a continuing interest in science.

Public recognition came to Lord Rayleigh in full measure; in 1904 he received the Nobel Prize for his discovery of argon, and he was one of the first recipients of the Order of Merit in 1902. Many honorary degrees and awards from learned societies followed. His contemporaries and successors alike place Rayleigh in that great group of nineteenth-century physicists that made British science famous throughout the world. He was above all a modest man, with a great sense of hunour. When he received the Order of Merit from the Crown he is reported as saying:

the only merit of which he personally was conscious was that of having pleased himself with his studies, and any results that may have been due to his researches were owing to the fact that it had been a pleasure to become a physicist.

Max Planck (1858–1947)
The German physicist Max Planck was foremost in demonstrating around the year 1900 that energy, at atomic level, exists in indivisible packets or quanta. In the case of light the unit is the *photon*, the energy of which is directly proportional to the light frequency according to the law:

$$Photon\ energy\ E = Constant\ h \times frequency\ f.$$

The constant h is known as 'Planck's Constant'. This law, and the concept of energy quanta, have had a profound influence on the development of atomic physics. It is at the heart of the technological design of devices such as television camera and display tubes, and lasers and light amplifiers for optical telecommunication systems.

Max Planck was born in Kiel, Germany; he studied at Munich University and later in Berlin. His early work was in thermo-dynamics and on 'black body radiation', i.e. a body that radiates perfectly at all frequencies, work which led him to the principle of energy quanta.

For his work in the quantum field he was awarded the Nobel Prize in 1918.

His collaboration with Albert Einstein before and after the First World War made Berlin a world centre for the study of theoretical physics.

References

1   Rollo Appleyard, *Pioneers of Electrical Communication* (Macmillan and Co. London, 1930).
2   *Pioneers in Telecommunications*, British Telecom Education Service, BT HQ, 81 Newgate Street, London, 1985.
3   William Cramp, *Michael Faraday* (Sir Isaac Pitman and Sons Ltd., London, 1931); John Meurig Thomas, *Michael Faraday and The Royal Institution*, (Adam Hilger, Bristol, 1991).
4   D.K.C. Macdonald, *Faraday, Maxwell and Kelvin* (Heinemann, London, 1965).
5   A. Feldman and P. Ford, *Scientists and Inventors* (Aldus Books, London, 1979).
6   T. Hey and P. Walters, *The Quantum Universe* (Cambridge University Press, 1987).
7   R.B. Lindsay, *Lord Rayleigh: The Man and His Work* (Pergammon Press, 1970).

Further reading

J.G. O'Hara and W. Pricha, *Hertz and the Maxwellians*, IEE History of Technology Series, No. 8 (London, 1987).
P. Tunbridge, *Lord Kelvin: His Influence on Electrical Measurements and Units*, IEE History of Technology Series, No. 18 (London, 1982).
G.R.M. Garratt, *The Early History of Radio: From Faraday to Marconi*, IEE History of Technology Series. No. 20 (London, 1994).

Chapter 3

The first telegraph and cable engineers

3.1 The visual telegraph (Semaphore)

Claude Chappe (1763–1805)

The first practical system of long-distance telegraphic comunication was the semaphore which used manually movable arms or closable apertures mounted in towers, usually on hilltops and within line-of-sight of one another, to convey messages letter by letter. The French revolution stimulated the need for swift and reliable communication throughout France in the 1770s – a need met by a brilliant French engineer Claude Chappe. Eventually 500 semaphore stations, spanning some 5,000 km, had been installed throughout France.

News of Chappe's invention had reached the British Admiralty and stimulated Lord George Murray (1761–1803) to develop a visual telegraph system using holes closed manually by movable wooden shutters. A system of 15 stations was installed between London, Deal and Portsmouth – some of them on sites still known as 'Telegraph Hills'. The semaphore may be regarded as a precursor of present-day microwave radio-relay systems which often use hill-top sites.

However, the semaphore with its dependence on the transmission of light through the atmosphere was susceptible to severe disruption in bad weather. And the rate of transmission of information was measured in words (or 10s of bits) per minute, compared with a million bits per second or more on microwave systems. [1] [2]

3.2 The electric telegraph

The discoveries in electro-magnetism in the early 19th century – notably Volta's battery, Faraday and Oersted's discoveries on the relation between electricity and magnetism – created the possibility of the electric telegraph, one using the transmission of an electric current along a conducting wire. Unlike the

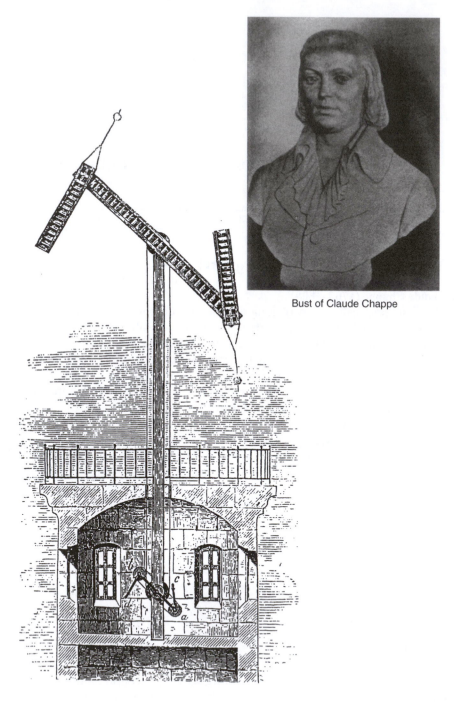

Bust of Claude Chappe

Figure 3.1   Chappe and semaphore
Showing movable arms and handles to operate them

Figure 3.2　Wheatstone and Cooke

semaphore, this clearly was immune to the effects of weather and offered speedier transmission of information.

Although there had been attempts to devise an electric telegraph from 1800 onwards in Britain, Russia, Bavaria and Switzerland, none of these found practical use on any scale. It was the partnership of Wheatstone and Cooke in the 1830s that led to the first electric telegraphs in the UK, inspired mainly by the need of the expanding railway system for a fast and reliable means of communication to ensure regular running of the trains, and especially standardised time-keeping throughout the country.

Charles Wheatstone (1802–1875)
William Fothergill Cooke (1806–1879)
Charles Wheatstone was born in 1802 in Gloucester into a family of musical instrument makers; his interest in music led him to experiment with sound and how it was produced. Other experiments he carried out included an unsuccessful attempt to measure the speed of propagation of an electrical current along a wire. For his experimental work he was elected a Fellow of the Royal Society in 1836.

W.F. Cooke was born four years after Wheatstone in 1806; his father was a surgeon and a professor of anatomy. His early life gave little indication of future scientific interests. After leaving university he became a soldier in the East India Company army, but retired after a few years on health grounds.

In 1836 Cooke attended some telegraphic experiments which Professor Munke at Heidelberg was making with a needle galvanometer, and was seized with the idea of turning this into a commercially practicable telegraph system. He consulted first Michael Faraday and then Professor Wheatstone at King's College, London who had also been conducting experiments with telegraphic apparatus. He built his own equipment and wrote a pamphlet about it, suggesting that telegraph wires could easily be laid along railways.

When Cooke and Wheatstone met in 1837, both were convinced of the possibilities of the telegraph and decided to form a partnership to develop and exploit it, taking out their first patent on 10 July 1837. This described a telegraph system using five magnetic needles, of which each could be deflected by an electric current as in Oersted's original experiment. By deflecting the needles successively in pairs any one of 20 letters of the alphabet could be selected in turn. The deflection of one needle could be used to point to a numeral. This system initially required five wire lines, later reduced to two by using a code.

The directors of the Great Western Railway were impressed by the invention and some 20 km of line were installed in 1838 between Paddington and West Drayton, using the Five Needle Telegraph.

By 1852, 4,000 miles of telegraph were in use in the UK. Cooke and Wheatstone continued to improve their telegraph system, eventually using only one needle and a signal code. There was, however, considerable ill-feeling between the two men as to who was the sole inventor of the electric telegraph. Perhaps the most appropriate comment is that both men were eventually knighted for their achievements in telegraphy.

Developments in electric telegraphy were also taking place in the USA of America, notably by S.F.B. Morse.

Samuel F.B. Morse (1791–1872)

Samuel Morse was born in 1791 at Charlestown, Massachusetts; he studied at Yale College, graduating in 1810. He had a talent for painting and studied in Europe, having a picture exhibited in the Royal Academy, London. He returned to the USA in 1815, but his hopes for a career in painting were only partially fulfilled by portraiture. In 1825 his wife and parents died and he had lost money in failed artistic ventures and paying debts incurred by his father.

However, he had learned about electricity at college and it had become a hobby for him in the 1820s. He attended lectures on electromagnetism in New York in 1826 and, in 1832 met a fellow passenger on a return trip from Europe, with whom he discussed and made notes on the possibilities of using an electromagnet as the receiving element in a telegraph system.

By 1835 he had made a printing telegraph in which a key could switch on the current to an electro-magnet as long as the key was held down, so causing a pencil to make long or short marks on a moving strip of paper.

Morse's key invention was undoubtedly the code that bears his name – the 'Morse Code'. He assigned different combinations of dots and dashes to the letters of the alphabet and the numerals. Experienced operators learned to 'read' morse by listening to the sounds of the clicking electro-magnet, without having to see the dots and dashes on the paper tape, thus creating the simplest of all telegraph systems using a key, battery, line and sounder.

The British 'needle telegraph' was later adapted to use the Morse Code by interpreting a right deflection as a 'dot' and a left as a 'dash'.

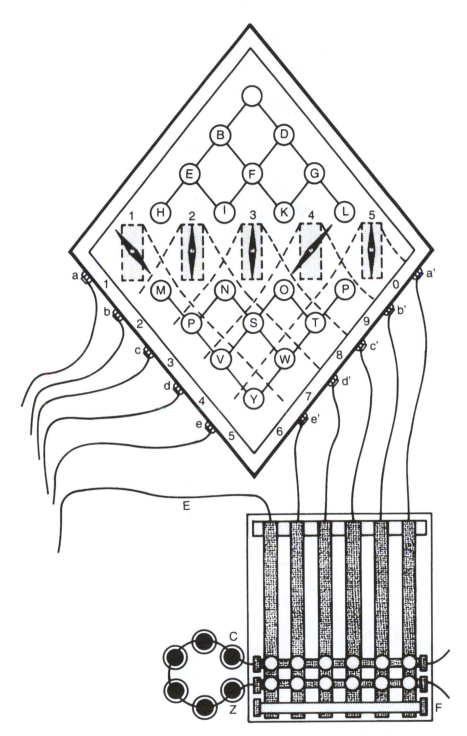

Figure 3.3   Five-needle telegraph

Figure 3.4   Morse

E	–	12,000
T	—	9,000
A	– —	8,000
I	– –	8,000
N	— –	8,000
O	– –	8,000
S	– – –	8,000
H	– – – –	6,400
R	– – –	6,200
D	— – –	4,400
L	—	4,000
U	– – —	3,400
C	– – –	3,000
M	— —	3,000
F	– – – –	2,300
W	– – —	2,000
Y	– – – –	2,000
G	— — –	1,700
P	– – – – –	1,700
B	— – – –	1,600
V	– – – —	1,200
K	— – —	800
Q	– – – —	500
J	– – – –	400
X	– – – – –	400
Z	– – – –	200

Figure 3.5   Morse Code

Morse's original code showing relation of lengths of dot and dash groups to the quantities of type found in a printer's office.

Morse had a difficult time launching his invention and he nearly starved until, in 1843, Congress finally gave him funds to set up a trial system between Baltimore and Washington in 1844. From that time telegraph systems expanded throughout the world and Morse eventually, after several lawsuits, became a wealthy man.

It is difficult to over-estimate the value and significance of the Morse Code. Not only did it provide the basis for a simple, robust and easily operated overland line telegraph system, it also served for point-to-point and mobile, e.g. ship-to-ship and ship-to-shore, radio communication. For such applications the ability of the Morse Code to be heard and interpreted by a skilled operator under conditions of high levels of radio interference and noise is of paramount importance, especially for distress calls. These are, no doubt, reasons why the Morse Code continued to be used throughout the world until 1993 when the International Maritime Organization (IMO) introduced a standardised Global Distress and Calling system for radio and satellite systems using digital technology.

But perhaps of even greater importance is the fact that the concept, inherent in the Morse Code – of transmission of information by coded groups of on/off signals – has led to a major revolution in communications: the use of 'digital technology' for the transmission and switching for all types of information, including audio, data, facsimile and television.

It is interesting that Morse's Code is efficient in terms of modern coding theory because the most probable symbols have the shortest codes.

By the time of Morse's death in 1872 other inventors and engineers had taken over from him, improving his system – Edison in America, Werner Siemens in Germany and his brother William in England.

The printing telegraph

David Edward Hughes
A London-born professor of music in Kentucky, David Edward Hughes, invented in 1854 a letter-printing telegraph with a 52-symbol keyboard in which each key caused the corresponding letter to be printed at the distant receiver. This was achieved by a rotating wheel bearing 28 letters of the alphabet and other signs on it, and a clutch mechanism actuated by an electro-magnet which brought the wheel momentarily to rest when the desired letter was above a moving strip of paper – an arrangement which had some similarities to the familiar type 'golf-ball' in a modern typewriter. When the sending operator pressed one of the keys on the keyboard, the electric impulse sent to line stopped the rotating type-wheel at just the right instant of time to cause the desired letter to be printed.

Hughes's achievement was all the greater because the typewriter had not been invented at that time! The teleprinter, the 'Telex' system, and present-day computer keyboards/visual displays are direct descendents of Hughes's invention.

3.3 Duplex, quadruplex and time-division multiplex telegraphy

As telegraph traffic grew, inventors sought means whereby each cable or wire could carry two or more messages simultaneously, thereby saving the cost of extra cables or wires – an important consideration on long-distance routes. The first to use a single telegraph wire for sending two messages simultaneously, one in each direction, was Dr Gintl in Vienna in 1853. His 'duplex-circuit' involved the use of an artificial line, simulating the real line, in a balanced-bridge arrangement, thereby enabling a message to be sent at the same time that one was being received, without mutual interference.

Thomas Alva Edison (1847–1931)
Edison, who had worked as a telegraph operator from the age of 15, turned his brilliantly inventive mind to the improvement of the telegraph. His patents included a 'duplex circuit' and, in 1874, a 'quadruplex circuit' enabling the simultaneous transmission of four messages on a single wire.

The next important development in telegraphy was the invention of 'time-division multiplexing' – attributed to Baudot.

Emile Baudot (1845–1903)
Emile Baudot was an officer in the French Telegraph Service who invented a system based on the use of a 'five-unit code', each letter of the alphabet being represented by a unique combination of the five elements. He combined the use of the five-unit code with time-division multiplexing – the latter being achieved by a rotating commutator at the receiving end of the line, synchronized with a similar commutator at the sending end. This enabled a single wire line to be used for the simultaneous transmission of as many messages as there were segments on each commutator. Each sending operator had a keyboard with five keys, corresponding to the five-unit code, whilst at the receiving end the message appeared on a strip of paper passing through a printer.

Baudot's system was first introduced in 1874 and officially adopted by the French Telegraph Service in 1877. His invention is particularly significant because it embodied two principles that were later to become of even greater importance:

(1) the representation, as with Morse, of a symbol by a code, and
(2) the concept of time-division multiplexing.

The first foreshadowed pulse-code modulation, while the second was a forerunner of modern digital techniques.

3.4 The first under-sea telegraph cables

Once over-land telegraph cables had been established in Great Britain and the Continental countries there was a clear need, for social, news, business and diplomatic purposes, to provide links across the English Channel.

The first Dover-Calais telegraph cable was laid in 1850 but proved vulnerable to damage by trawlers. A second cable, with four copper conductors insulated with gutta-percha and protected by tarred hemp and armouring wires, was laid in the following year and proved more successful, remaining in operation for many years.

This success stimulated the provision of telegraph cables across the North Sea and in other parts of the world, e.g. Italy to Corsica and Sardinia, India to Ceylon and Tasmania to Australia. However, although land links to British possessions in the Middle and Far East were possible, these were liable both to technical breakdown and political disruption or interception in times of trouble. Thus for strategic as well as practical considerations, an under-sea telegraph cable system from the UK via Gibraltar, the Mediterranean, the Red Sea to Singapore, China and Australia was projected.

The initiative taken by Great Britain in setting up a world-wide telegraphic network prompted Reuter to set up his headquarters in London, which became established as the world's first telegraphic news and financial information distribution centre – an early version on a small scale of today's Internet! There remained, however, the most important cable yet to be laid, one to span the Atlantic ocean.

3.5 The first trans-Atlantic telegraph cable

Cyrus W. Field (1819–1892)
William Thomson (1824–1907)
The story of the first trans-Atlantic telegraph cable is an epic of courage, enterprise and perseverance. It owes much to a great American citizen, Cyrus W. Field, whose untiring efforts provided a major impetus throughout the project. But equally important was the scientific advice given by Professor William Thomson (later Lord Kelvin).

The first attempts in 1857 to lay the cable between Valentia, Ireland and Newfoundland by British and American warships were unsuccessful due to breaks in the cable caused partly by unsatisfactory cable-laying techniques.

An attempt in 1858 was more successful and on 14 August 1858 Queen Victoria and the American President exchanged messages of congratulation. But within a month of the first messages the cable failed due to an insulation breakdown caused by an over-zealous operator who had applied an excessive voltage in an effort to improve the signals.

The American Civil War delayed work on the cable for a time, but by 1865 the Atlantic Telegraph Company had raised new capital, redesigned the cable and chartered Brunel's *Great Eastern* with improved laying equipment. The *Great Eastern* was a huge passenger liner, unsuccessful as such, but was the only ship afloat with capacity to hold the entire 3,700 km of cable.

Professor William Thomson, see Chapter 2, had been engaged as a consultant

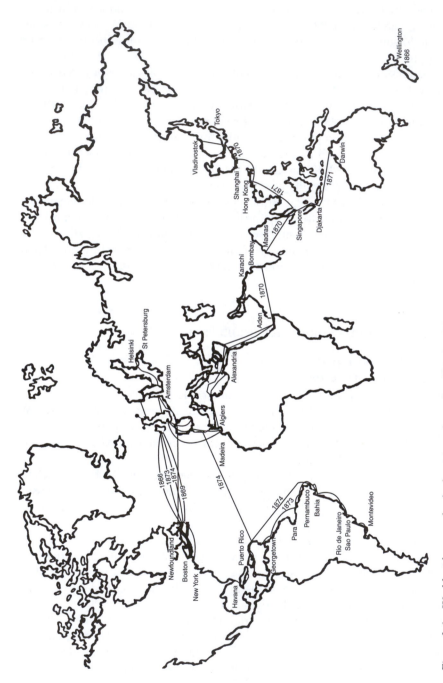

Figure 3.6   World-wide network of submarine telegraph cables (1875)

to advise and report on the electrical performance of the cable during the design stage and on board the *Great Eastern* during the laying operations.

A long cable behaves as large distributed capacitance and this severely limits the rate at which the signal current waveform can rise or fall, and therefore restricts the rate at which messages can be sent over the cable. This in turn decides the earning capacity of the cable and the ultimate financial return to the cable company. The theoretical prediction of the electrical performance owes much to the work of Kelvin and Heaviside, as does the design of the sending and receiving equipment for optimum performance. Kelvin's invention of the highly sensitive mirror galvanometer was of critical importance in the operation of the cable.

The first of the renewed attempts in 1865 to lay the cable was a failure due to a breakage and the *Great Eastern* had to return to Valentia. A second attempt in 1866 was successful and on 27 July of that year the first telegraph messages were transmitted between the UK and the USA of America.

Not only was the first successful trans-Atlantic telegraph cable of great importance as the only telecommunication channel then available between the two countries, it also gave valuable practical experience in overcoming the difficult problems of laying and recovering cables in the great depths of the Atlantic Ocean.

This experience, and the confidence it generated, were of immense value when, nearly a century later, the first trans-Atlantic telephone cable came to be laid.

By 1875 a world-wide network of telegraph cables had been laid, linking the UK, the USA, India, the Far East and Australia – a tribute to the Victorian telegraph pioneers of England and America. [3]

References

1 *Pioneers in Telecomunications* (British Telecom Education Service, BT HQ, 81 Newgate Street, London, 1985).
2 *From Semaphore to Satellite* (The International Telecommunication Union, Geneva, 1965).
3 Bernard S. Finn, *Submarine Telegraphy: The Grand Victorian Strategy* (Science Museum, London, 1973).

Chapter 4

The first telephone engineers

4.1 The telegraph learns to talk: the invention of the telephone

It may well have been Alexander Graham Bell's successful demonstration on 10 March 1876 of what was later claimed to be the world's first demonstration of the electronic transmission of intelligible speech – enshrined in Bell's historic call to his assistant 'Mr Watson, come here, I want to see you' – that set in motion what eventually became a major step forward in communication at a distance. However, the history of the invention of the telephone is complex and Bell's was by no means the only contribution.

The need for the telephone, as compared with the telegraph, is not difficult to find. The world of the 19th century was that of the Industrial Revolution; science, technology, manufacture and trade were on the march, and new political forces were making themselves felt. There was a growing need to communicate, rapidly and in the most natural way possible. The electric telegraph met these requirements only partially, it was slow and conveyed little of the personality of the communicators.

Human speech, developed during the many millennia of Man's evolution and closely linked to the maximum rate at which he could create, absorb and respond to information, was clearly the most natural and useful mode of person-to-person communication. Fortunately the discoveries in the field of electromagnetism made by Oersted, Faraday, Henry and others in the early 19th century had not only made possible the electric telegraph, they also created the foundation of scientific knowledge necessary for the invention of the telephone. Furthermore, by 1875 there was already in existence an extensive and growing network of intercity and international telegraph wires and cables that must have prompted the idea that if these could carry the electric currents that corresponded to telegraph signals, might they not be adapted to carry electric signals corresponding to speech?

Alexander Graham Bell (1847–1922)

Alexander Graham Bell was born in Edinburgh in 1847. Both his father and grandfather were interested in the nature of speech and the human vocal system, this no doubt stimulated a similar interest in Alexander. The family moved to Canada and later to Boston where Bell became Professor of Vocal Physiology in Boston University.

He was a succesful teacher of deaf children, making a notable contribution by disproving an earlier contention that the deaf could not be taught to speak. Besides teaching the deaf and dumb Bell was absorbed in the scientific study of sound and the possibilities of transmitting it by electricity or light. [1]

Bell's telephone comprised a transmitter and a receiver each with a thin iron diaphragm in front of an iron core surrounded by a coil of wire, the iron core in the transmitter being magnetised by direct current in the coil from a battery.

Figure 4.1   Bell and Edison

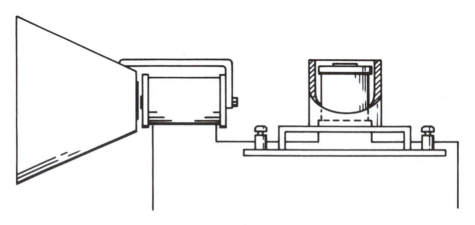

Figure 4.2   Bell's electromagnetic telephone transmitter and receiver

Sound waves falling on the transmitter diaphragm caused it to vibrate and generate a similar variation of the magnetic field. This in turn induced an undulating electric potential and current in the coil – a direct application of Faraday's principle of electromagnetic induction. The undulating current from the transmitter was transmitted over conducting wires to the coil in the distant receiver where it generated a varying magnetic attraction on the diaphragm, thus producing a copy of the original sound waves.

This is the arrangement described in Bell's first patent No. 174,465 of March, 1876. His second patent, No.186,787 of January, 1877 described the use of a permanent magnet, enabling the battery to be dispensed with. This patent included, almost as an afterthought, and coincident to a caveat entered by a fellow inventor Elisha Gray, a reference to a liquid variable resistance transmitter.

It is remarkable that most modern telephone receivers use principles similar to those established by Bell in 1876. However, a marked improvement in the sensitivity of the transmitter was achieved, perhaps at some expense of sound quality and long-term reliability, by Thomas Alva Edison's invention of the carbon granule transmitter, the resistance of which varied in sympathy with the sound wave pressure on it. This, until recently, was used in most telephone systems. Edison's invention of the transformer, comprising two coils of wire on an iron core, much as in one of Faraday's experiments, further improved the telephone by enabling the battery circuit to be separated from the transmission line, and by providing a better impedance match to the line.

The telephone was first shown to the public at the USA 100th Anniversary of

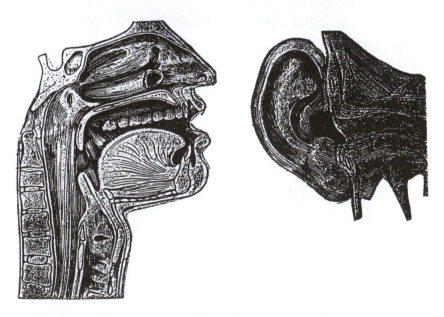

Figure 4.3   The world's most advanced form of communication – human speech

Figure 4.4   Alexander Graham Bell, inventor of the telephone (US Library of Congress)

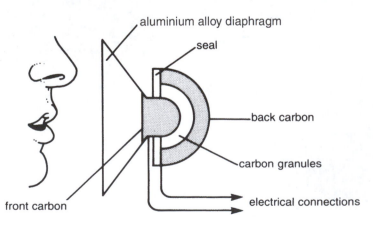

Figure 4.5   Edison's carbon granule transmitter

Independence Exhibition in Philadelphia in 1876, where the Emperor of Brazil was astounded to hear Bell's voice, exclaiming 'it talks!'. It was also demonstrated to Queen Victoria at Osborne, England in 1877, who is said to have been 'most impressed'.

4.1.1 The telephone age had begun

Bell seems to have been modest about his achievement and is on record as saying:

. . . if he had known more about electricity he would not have invented the principle of the telephone.

However, his words on 10 March, 1876 were nevertheless prophetic and showed he had a fair vision of what the telephone might do:

This is a great day with me and I feel I have at last struck the solution of a great problem and the day is coming when telephone wires will be laid on to houses, just like water or gas, and friends will converse without leaving home.

He foresaw too that:

. . . the telephone would be a new factor in the new urbanization, without the telephone the 20th century metropolis would have been stunted by congestion and slowed to the primordial pace of messengers and postmen. And the modern industrial age would have been born with cerebral palsy.

4.2 The other inventors of the telephone

The history of the invention of the telephone is complex, confused and incomplete. Elisha Gray, whose own contribution was by no means insubstantial and who sought to challenge the Bell patents, wrote:

The history of the telephone will never be fully written. It is partly hidden in 20 or 30 thousand pages of testimony and partly lying in the hearts and consciences of a few whose lips are sealed some in death and others by a golden clasp whose grip is tighter.

It is clear from the literature that Bell's contribution was by no means the unique and original one that some of his supporters have claimed. Between 1820 and 1860 a score or more of inventors had described, and some had made to work, devices that in part anticipated the essential features of Bell's telephone. These inventors included Meucci (Italy), Bourseul (France), Reis (Germany), Elisha Gray and Thomas Edison (USA). [2]

With the growing commercial value and importance of the telephone, and the massive financial issues that were involved, it is hardly surprising that the 20 years following the issue of Bell's patent saw more than 600 cases of patent litigation, including a long drawn-out but unsuccessful suit by the USA Courts against the Bell Telephone Company's monopoly claim to the use of electricity for the transmission of speech. However, in every case Bell was upheld as the inventor of the telephone, perhaps in no small measure because his demonstration of 1876 clearly revealed its capability for the transmission of good quality, intelligible and recognisable speech.

4.3 The social and economic value of the telephone

Not all in the 19th century were convinced of the value and future prospects of the telephone. The President of the mighty Western Union Telegraph Company in the USA turned down the opportunity to buy the rights to Bell's patents for a song, thereby inhibiting indefinitely the possibility of a rational and economic integration of the telegraph and telephone services in that country.

Some of the users were not altogether satisfied with the performance of the new device. In 1880 Mark Twain, who was evidently having difficulty in hearing over the lines of the Hartford Telephone Company, wrote in a letter to the *New York World* newspaper:

It is my heart-warm and world-embracing Christmas hope and aspiration for all of us, the high, the low, the poor, the rich, the admired, and the despised, may eventually be gathered together in a heaven of everlasting rest and peace and bliss except the inventor of the telephone!

Others have objected to the telephone for its intrusions into personal privacy; Bell himself refused to have one in his study! Some have criticised the telephone for its adverse effect on letter-writing, a civilised and cultivated art in Victorian times. The speed and immediacy of the telephone are not without disadvantages; for example, the instability and collapse of the Wall Street stock market in 1929 have been attributed in part to the use of the telephone for the panic selling of shares. This has an interesting parallel with the effect of computer-programmed share selling today. And many a military commander in the field has resented 'being at the end of a telephone'.

But the overall evidence in favour of the telephone is clear and overwhelming since more than 1,000 million are now in use in the world.

The value of personal identification provided by the telephone is obvious and far reaching in many areas of human life, ranging from a young child's first call to a grand-parent, to a call for help by a potential suicide on a 'Samaritan' telephone, and to a Presidential call on an international 'hot line' heading off a nuclear disaster. As the American sociologist Sydney H. Aronson has written:

... it has helped to transform life in cities and on farms, and to change the conduct of business ... it imparted an impulse towards the development of a 'mass culture' and a 'mass society', at the same time it affected institution patterns in education and medicine, in law and warfare, in manners and morals, in crime and police work, in the handling of crises and the ordinary routines of life. It markedly affected the gathering of news and the patterns of leisure activity, it changed the context and even the meaning of neighbourhood and of friendship, it gave the family an important means to adapt itself to the demands of modernization and it paved the way both technologically and psychologically for the 20th century media of communication: radio and television broadcasting. [3]

One may be justified in feeling that this was a remarkable outcome from an invention that was in itself of quite remarkable simplicity.

4.4 The beginning of automatic telephone exchange switching

The problem of connecting a calling to a called customer was at first solved by manually operated telephone exchanges in which an operator simply plugged in a cord between the corresponding incoming and outgoing telephone line terminals on a switchboard.

This system had the advantage that it provided, from the customer viewpoint, good service since the operator could, in systems with small numbers of lines, readily find the called customer by name, and answer queries made by the caller. But it became cumbersome when large numbers of lines were involved, a difficulty only partially solved by the use of multiple switchboards with groups of operators. And there was the inherent problem of overhearing the customers' telephone conversations by the operators and the consequent lack of privacy.

Almon Brown Strowger (1839–1902)

The lack of privacy was the motivation that in 1889 led a Kansas City undertaker, Almon Brown Strowger, to seek a solution to the interconnection problem by inventing an automatic telephone switching system that dispensed with operators.

Strowger christened his system 'the girl-less, cuss-less telephone' – the latter adjective presumably being justified because it eliminated the delays occasioned by inattentive operators!

It has been said that Strowger found that he was losing money in his undertaking business because a switchboard operator at the Kansas City telephone exchange was married to a rival undertaker and she connected callers making funeral arrangements to her husband instead of Strowger!

Strowger's invention, which he first modelled with a collar box and matches to represent telephone lines, was of remarkable simplicity.

It comprised two basic elements:

(a) a device for use by the customer which created trains of on-off pulses of current corresponding to the digits 0 to 10 (this eventually became the familiar circular 10-hole telephone dial), and

(b) a switch at the telephone exchange in which a rotating arm was caused to move step-by-step over a semi-circular arc of 10 contacts, each contact being connected to a customer's line, the stepping motion being controlled by the pulses of current on the calling customer's line via an electro-magnet.

Almon Brown Strowger (1839–1902)

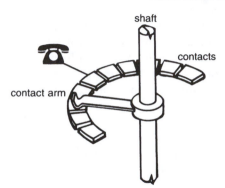

(a) Single-level switch

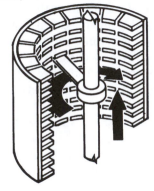

(b) 10-level switch

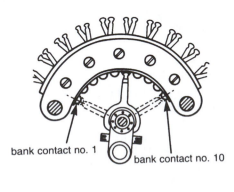

(c) Modern Strowger switch

Figure 4.6   Strowger automatic telephone switching system

In this form only 10 lines could be catered for; by providing 10 such arcs of 10 contacts, one above the other, and arranging that the rotating arm stepped first vertically and then horizontally, 100 lines could be accommodated. The next step was to increase the horizontal arc to 20 contacts, accommodating 200 lines. Eventually Strowger's competitors developed step-by-step switches with 100 contacts in each horizontal arc, making possible 1,000-line exchanges. Beyond this number of lines Strowger's original concept became impracticable and uneconomic. New approaches to exchange design were developed in which a Strowger-type switch (a line-finder) first found the calling customer's line and then passed the call on to other switches which could be shared with many customers, thereby economising in the amount of equipment required and making possible 100,000-line exchanges, suitable for large cities. Strowger's US patent was filed in March 1889 and issued in May 1891; the first practical application was at La Porte, Indiana in 1892. The first Strowger exchange in the UK was opened at Epsom, Surrey in 1912. By 1918 it had become the norm for automatic exchanges in the UK.

As use of the telephone expanded, and the numbers of digits to be dialled grew, calls had to be routed via two or more exchanges. This stimulated further development in the technique of switching called 'signalling' by which the dialled digits could be stored at the first exchange and re-processed for onward transmission to the following exchanges. [4]

Strowger lacked the engineering knowledge fully to develop and exploit his invention but with associates the Automatic Electric Company was formed. He died in 1902 but the company continued with the further development and large-scale manufacture of automatic switching equipment based on the Strowger switch continued well into the 1970s.

The Bell Telephone System was at first slow to adopt Strowger's switch; perhaps the 'not invented here' syndrome may have been in part responsible but when convinced of the economic advantages of automatic working, and satisfied that satisfactory quality of service was achievable, Bell began to buy in quantity from the Automatic Electric Company.

In 1916 Western Electric, Bell's manufacturing arm, bought manufacturing rights for Strowger switches and commenced its own production in 1926. [5]

From the USA the Strowger automatic switching system spread all over the world, and although several other systems of automatic switching based on electro-mechanical techniques were developed in the USA and Europe, up to the 1970s the Strowger system was still the most widely used. [6]

References

1 Robert V. Bruce, *Alexander Graham Bell and the Conquest of Solitude* (Victor Gallancz, 1973).
2 William Aitken, *Who Invented the Telephone?* (Blackie and Son Ltd., 1939).

3 Sydney H. Aronson, 'The Sociology of the Telephone', *Intl. Journal of Comparative Sociology*, Vol. XII, No.3, 1971.
4 S. Welch, *Signalling in Telecommunication Networks*, Institution of Electrical Engineers, Peter Peregrinus, 1981).
5 *A History of Engineering and Science in the Bell System: The Early Years 1875–1925*), Chapter 6, (Bell Telephone Laboratories, 1975).
6 R.J. Chapuis, *100 Years of Telephone Switching, 1878–1978* (North Holland Publishing Co., 1982).

Chapter 5

Inventors of the thermionic valve

5.1 The challenge of long-distance wire and radio communication

As the telephone line networks in America and Europe expanded to include inter-city transmission over ever longer distances, so problems of unsatisfactory reception due to attenuation of the line current and poor signal-to-noise ratio began to be apparent. At first these difficulties were fought by using heavier weight, and therefore lower attenuation, copper conductors, but this approach became impracticable and uneconomic beyond a certain distance. There was also the problem of poor transmission quality due to the greater attenuation of the higher frequency components of the transmitted voice signals compared with the low frequencies.

The ideas initiated by Heaviside on inductive loading of telephone cables, developed by Michael Pupin and by George Campbell of the ATT Co.mpany, improved transmission quality but left unsolved the problem of overall line loss. Early attempts at a repeater/amplifier involved mechanically coupling an electromagnetic receiver and a carbon-granule microphone in a battery circuit; a device which produced amplification but yielded poor quality. It was clear that what was needed was some means of amplifying the voice signals but without distorting them.

And as radio had evolved beyond the sending of Morse code telegraph signals by spark transmitters to voice transmission, so the need arose for means for generating continuous radio-frequency carrier waves and modulating voice signals on to them – techniques that were also essential for multi-channel frequency-division wire and cable carrier systems.

The technique of 'modulation' of a carrier wave involved varying its amplitude or frequency in response to a voice or video signal or train of coded pulses; the later mode became the key to today's 'digital technology' increasingly used in telecommunications and broadcasting. 'Frequency division' involves the use of different frequencies for carrier waves, often adjacent to one another in the frequency spectrum, bearing different signals.

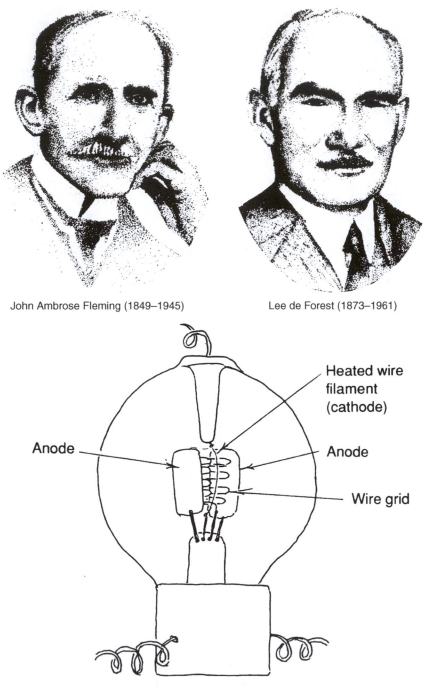

John Ambrose Fleming (1849–1945)   Lee de Forest (1873–1961)

Heated wire
filament
(cathode)

Anode   Anode

Wire grid

Lee de Forest's 'Audion'

Figure 5.1   Fleming, de Forest and audion valve

The pioneering work needed to create a device that could amplify, generate and modulate was carried out largely by three men – an Englishman, an Austrian and an American. The device they evolved was the 'thermionic valve', so called from the Greek *thermos*, meaning warm, which referred to the generation of electrons by a heated wire. American terminology referred to it, perhaps less descriptively, as a 'vacuum tube'.

The thermionic valve, which was eventually used in millions on a world-wide scale in telecommunications and broadcasting, held the field from the early 1900s until the arrival of the transistor in the 1950s.

John Ambrose Fleming (1849–1945)
John Ambrose Fleming was born in Lancaster in 1849. He graduated from University College, London in 1870 and entered Cambridge University in 1877 where he worked on electrical experiments under James Clerk Maxwell in the Cavendish Laboratory. He later became Professor of Electrical Engineering at University College, where he carried out experiments in wireless telegraphy and co-operated with Marconi in the design of a transmitter for the first radio transmissions across the Atlantic in 1901. [1] [2]

Fleming discovered in 1904 that the electron flow in a vacuum tube with a heated wire cathode was uni-directional towards an anode held at a positive potential relative to the cathode. He realised that such a device could be used to generate a direct current when an alternating potential was included in the circuit, that is to say it would act as a 'rectifier'. Furthermore, this action was effective even for the high-frequency signals involved in radio waves, and could be used as a 'detector' enabling a telegraph or voice signal to be extracted from a modulated radio carrier wave. As such it was a more stable and efficient device than the crystal detectors then in use. Although Edison as long ago as 1883 had observed a similar effect involving electrons, it fell to Fleming to realise its potential for detection and to apply it effectively.

Fleming's device became a 'diode' thermionic valve, since it had two electrodes. The Fleming diode could not act as a generator or amplifier of electrical oscillations; its function was essentially that of a detector in a radio receiver or a rectifier for the generation of direct current from an alternating current power supply. But the Fleming diode served another valuable purpose, it triggered the next and most important development of the thermionic valve that enabled it to amplify and generate oscillations.

In 1929 Fleming was knighted for his pioneering work in wireless telegraphy, of which the diode was a part; he lived to the age of 96 and was able to see much of his work come to fruition and wide application in the course of his lifetime.

Robert von Lieben (1878–1914)
Robert von Lieben, an Austrian physicist working in Vienna on the amplification of telephone signals, made the next important step forward. In 1906 he added a third electrode to Fleming's diode, in the form of a wire mesh 'grid' between the electron emitting cathode and the positive potential anode, his

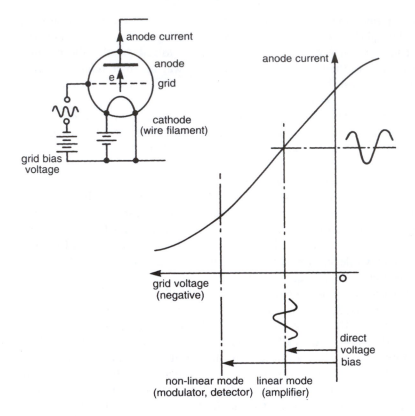

Figure 5.2   Anode current/grid voltage characteristics of the triode thermionic valve

device then becoming a 'triode'. By varying the negative potential of the grid the current from the cathode to the anode could be varied, and if an alternating signal voltage was impressed on the grid the anode current varied correspondingly and an amplified signal voltage appeared across a resistance in the anode circuit.

This invention was clearly of major importance, opening the door as it did not only to amplification, but also to the generation of stable high-frequency oscillations by coupling together circuits connected to the anode and the grid. However, von Lieben's work was not fully exploited, perhaps because of his early death at the outbreak of war in 1914, and it was an American Lee de Forest to whom most of the credit for the development of the triode thermionic valve should go.

Lee de Forest (1873–1961)
Lee de Forest was born in 1873 at Council Bluffs, Iowa, USA. He initially studied mechanical engineering at Yale University but later became interested in radio waves and the new developments in wireless telegraphy, receiving a Ph.D. in 1899 for his thesis on this subject. In 1901 he devised, in competition with

Marconi, means for speeding up the sending of wireless telegraphy signals and was able to interest the US Army and Navy in his work which found application to news reporting in the Russo-Japanese war of 1904–5. He was a prolific inventor, being granted more than 300 patents, but his US Patent No. 841, 387 for a 'Device for Amplifying Feeble Electric Currents', issued in January 1907, was the most significant. It closely paralleled von Lieben's 1906 discovery in its use of a control grid; however in de Forest's 1907 patent this was to one side of, rather than across, the electron stream from the cathode, a limitation removed in his 1908 patent on 'Space Telegraphy' which showed the grid as a wire mesh through which the electrons flowed on their way to the anode, thus giving more effective control. This device he called an 'audion'. [3]

At first the de Forest audion was looked on as a very sensitive detector of radio waves, and – perhaps because of the inventor's pre-occupation with wireless telegraphy – its potential applications as an amplifier and oscillator were for some five years overlooked, 'even by the inventor'!. [4] It was John Stone Stone, who worked in the Boston Laboratory of American Bell, who realised the potential of the audion as an amplifier in the growing telephone network. Applications as modulators and generators of carrier waves followed, notably by engineers of the Bell System, including H.D. Arnold, R.V.L. Hartley, E.H. Colpitts and R.A. Heising; these are described in H.J. van der Bijl's book *The Thermionic Vacuum Tube and its Applications*. [5]

Eventually de Forest sold his 'glass bottle full of nothing!' to the ATT Co.mpany for $390,000 – a vast sum in those days but a clear indication of the value of this pioneering achievement. Lee Forest was a man of many parts – with a strong interest in cultural matters and music. He staged the first radio broadcast in history from the Metropolitan Opera House, New York in 1910 and in 1916 set up his own radio station, broadcasting news as well as other programmes. In 1923 he became interested in sound motion pictures and demonstrated one of the first 'talkies', doubtless using his audion as an amplifier!

His life, before the success of his audion, was eventful and hazardous – he, like many inventors, was not a very successful business man and he lost fortunes as well as making them. He was even placed under arrest at one time for allegedly attempting to use the mail to defraud – this in an effort to finance one of his many inventions.

From these early beginnings many improvements in the design of thermionic valves followed, to which the fundamental studies in thermionic emission from hot cathodes by O.W. Richardson (c.1901) provided a scientific base. Improved vacuum techniques enabled residual gases to be removed from within the glass envelope and stabilised performance of the valve, whilst the introduction of indirectly-heated cathodes by Whenelt (c.1904) gave greater freedom of circuit design. E.H. Armstrong, see Chapter 8, was one of the first to use the thermionic valve as a circuit element providing amplification, with positive feedback to increase the gain and, where needed, generate oscillations.

The invention of the thermionic valve was the essential key that opened the door to a wide range of developments in telecommunications and broadcasting.

Not only did it overcome the problem of distance, it made possible multi-channel carrier cable and radio systems that provided economically for the vast growth of inter-city telegraph and telephone traffic that began in the early 1900s. Without the thermionic valve sound and television broadcasting may well have been delayed until the advent of the transistor in the 1950s. And even today the thermionic valve has a continuing role in the high-power stages of broadcasting transmitters.

References

1 J.A. Fleming, *The Principles of Electric-wave Telegraphy* (Longmans Green Ltd., 1906).
2 Ambrose Fleming, *Memoirs of a Scientific Life* (Marshall, Morgan and Scott, London and Edinburgh, 1934).
3 *Pioneers in Telecommunications* (British Telecom Education Service, BT HQ, 81 Newgate Street, London, 1985).
4 *A History of Science and Engineering in the Bell System: The Early Years (1875–1925)*, Chapter 4, p. 256 (Bell Telephone Laboratories, 1975).
5 Van der Bijl, *The Thermionic Vacuum Tube and Its Applications* (ISBN 017720), published 1920. (Copy available from the Library of the Institution of Electrical Engineers, London.)

Chapter 6

The telegraph-telephone frequency-division multiplex transmission engineers

6.1 The demands of long-distance traffic growth and the solution

Just as the growth of telegraph traffic and the pressure of economic factors led the early telegraph engineers to develop duplex and quadruplex telegraph systems capable of transmitting two or four messages simultaneously on a single wire, so the growth of inter-city telephone communication created demands for providing ever more telephone channels, at first on open-wire lines and balanced pair cables. This was paralleled by the need similarly to provide more than just two or four telegraph channels on a single transmission path. And when coaxial cables became available these too were required to provide many telephone channels on each cable.

The solution to this problem came with the evolution of the concept of 'frequency-division multiplexing' in which each telegraph or telephone signal is modulated on a carrier wave of different frequency, transmitted over a common transmission path, selected at the receiving end by a 'wave filter' responsive only to a limited band of frequencies, and the telegraph or telephone signal recovered after demodulation of the carrier wave. An early approach to this concept was Alexander Graham Bell's 'harmonic telegraph' (US patents No. 161,739 of 1875 and 174,465 of 1876) in which vibrating reeds at the sending end, each tuned to a different frequency, pulsed an electric line current at the same frequency. This line current energised a similarly tuned reed at the receiving end. By such means Bell hoped to transmit several telegraph messages simultaneously over the same line, using what was a primitive form of frequency-division multiplexing based on mechanical, rather than electrical, resonance.

Similar ideas using mechanical resonance were suggested in the 1870s and 1880s by Gray, van Rysselbergh, Edison and others. However, the most significant advances came when electrical resonance, using tuned circuits with inductance and capacitance, was proposed in the 1890s for multi-channel telephony by

Michael Pupin (at Columbia University), and John Stone Stone in the USA, and by *Hutin* and *Leblanc* in France.

James Clerk Maxwell in England had noted, as early as 1868, the analogy between electrical resonance in a circuit containing inductance and capacitance and the then already well-known mechanical resonance in a system involving a mechanical mass and a spring.

The basic elements of a carrier telephony system involved:

(a)   a source of stable, continuous high-frequency electromagnetic oscillations, i.e. a carrier wave;

(b)   means for varying the amplitude of the carrier wave in sympathy with the amplitude of a voice wave or telegraph signal, i.e. *modulation*;

(c)   a wave-filter capable of selecting the modulated carrier wave from others of different frequency;

(d)   means for extracting the desired voice or telegraph signal from the modulated carrier wave, i.e. *demodulation*.

The triode valve, derived from de Forest's 'audion', not only fulfilled the function of a line amplifier by operating on the linear portion of its grid-voltage/anode-current characteristic, the non-linear portion enabled it to be used as a modulator or a demodulator, as was demonstrated by Van der Bijl and Heising of the Western Electric Engineering Department in America. [1]

The later addition of a second grid between the control grid and the anode further improved the performance of the valve by removing unwanted capacitance between the control grid and anode that could otherwise cause unwanted oscillations, thereby inhibiting its use as a high-frequency carrier amplifier.

The development of the thermionic valve received an early impetus from wireless telegraphy and telephony, where it replaced cumbersome and inefficient spark generators and high-frequency alternators, and provided a more sensitive and stable alternative to crystal detectors. In fact the first multi-channel carrier telephony systems on wire lines were sometimes referred to as 'wired wireless' systems.

Although modulation and demodulation in wire and cable carrier systems were later carried out by copper-oxide rectifiers, the thermionic valve remained in use in audio-frequency and carrier-frequency amplifiers until the advent of the transistor in the 1950s.

By 1912 John Stone Stone, an independent worker formerly with American Bell, had written: [2]

A new art has been born to us. The new art of high-frequency multiplex telegraphy and telephony is the latest addition to our brood of young electric arts.

But much remained to be done before the system design and technology were adequate for commercial use. The next two important and innovative steps in the development of frequency-division multiplex carrier systems arose from the theoretical and mathematical studies of engineers in the laboratories of Western Electric and the ATT Co.mpany in about 1914.

One was the creation of a modulation theory that demonstrated quantitatively the characteristics of the sidebands of an amplitude modulated carrier wave, and the second the principles on which could be designed wave-filters sufficiently selective to separate the sidebands.

6.2 The sidebands are revealed

J.R. Carson

It seems remarkable now that in the 1920s there were some, including the eminent scientist Sir Ambrose Fleming, who doubted the objective existence of the sidebands of a modulated carrier wave, regarding them as a convenient mathematical fiction and prefering instead to use the 'decrement' (or damping factor, determined by the ratio of the circuit's reactance to its circuit resistance and defined by its Q factor), which determined its response to a modulated carrier wave.

The first clue came in 1914 from the laboratory notebook of a young Western Electric engineer, C.R. Englund, which showed the physical relationship of the sidebands and the carrier of an amplitude modulated wave. [3]

But it was the theoretical analysis of J.R. Carson in 1915 that gave mathematical precision to this concept and revealed possibilities for major improvements in the carrier multiplex transmission of telephony. He showed that if a carrier of frequency f and a voice wave of frequency v were passed through a non-linear device (e.g. a valve operated on a curved part of its grid voltage/anode current characteristic), the output contained not only the original frequencies f and v but also an upper sideband f + v, and a lower sideband f − v. He also demonstrated that, if the carrier and its sidebands were passed in to another non-linear device, the output of the latter contained the original voice frequency v, together with higher frequency unwanted components that could be removed by a wave-filter.

Another important deduction from Carson's analysis was that it was sufficient to transmit only one of the two sidebands to achieve satisfactory voice quality, the carrier suppressed before transmission being replaced by another of the same frequency (or nearly the same within a few cycles per second).

This became known as the 'suppressed-carrier, single-sideband' mode of transmission. Since the frequency band occupied by the transmitted sideband could be determined by choice of the carrier frequency a number of sidebands, each corresponding to a different voice channel, could be stacked side-by-side in the wide frequency band available on the wire or cable transmission path.

In order to suppress the carrier a balanced modulator could be used, but the selection of a desired sideband required more than a simple tuned circuit; a 'wave filter' providing a uniform response over the frequencies of the sideband and rejection of other frequencies was needed. Wave filters were also required to extract a desired sideband from the assembly of sidebands – the 'multiplex' – at the receiving end of the line. Here too, engineers and mathematicians of Western Electric and the ATT Co.mpany provided solutions [1].

6.3 The sidebands are selected

G.A. Campbell and O.J. Zobel

G.A. Campbell had worked on the inductive loading of transmission lines to improve their frequency response and in 1910 turned his attention to the design of electric wave filters. using inductors and capacitors in various circuit configurations he devised low-pass, high-pass and band-pass filters that could accept frequencies in a pass range and reject to any desired degree frequencies outside that range.

The sharpness of transition between the pass range and the reject range, and the degree of rejection outside the pass range could be controlled by suitable circuit design, the circuit requiring more inductor and capacitor elements as the desired performance became more stringent. For voice transmission the pass range of frequencies transmitted ranged between about 200 and 2,500 hertz in early carrier systems; this was eventually increased to 300 to 3,400 hertz with 4,000 hertz spacing between the virtual carrier frequencies and became a world standard of the International Telecommunication Union.

O.J. Zobel joined the Engineering Department of the ATT Co., with a doctorate from the University of Wisconsin. His mathematical skill, combined with outstanding engineering insight, brought the highly esoteric art of wave filter design to new standards of performance. By 1920 he had evolved wave filter design techniques that improved the impedance/frequency characteristic within the pass-band, thereby minimising reflections, and enabled high peaks of attenuation (rejection of unwanted frequencies) to be provided outside but close to the pass-band, thus sharpening the transition region [4].

These filter developments, to which R.S. Hoyt of ATT Co. also contributed, had substantial economic value by enabling more single-sideband voice channels to be packed into the frequency band available on a wire or cable system.

6.4 Benefits of the single-sideband, suppressed carrier mode

The effects of Carson's work on modulation theory were both profound and widespread. One of the first applications of single-sideband working was to the long-wave 60 kHz trans-Atlantic radio telephone system in 1926, where the frequency space available in that part of the radio spectrum was severely limited. The absence of a transmitted carrier enabled the whole of the transmitter output power to be concentrated in the sideband, thereby achieving a substantial improvement in signal-to-noise ratio at the receiver.

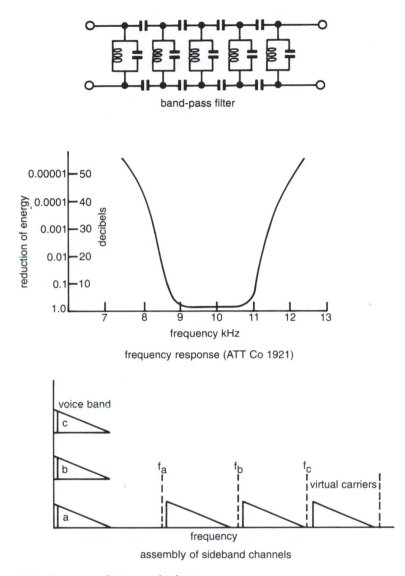

band-pass filter

frequency response (ATT Co 1921)

assembly of sideband channels

Figure 6.1   Frequency-division multiplexing

The single-sideband, suppressed-carrier, mode conferred substantial advantages in multiplex carrier systems. It doubled the number of voice channels that could be accommodated in a given frequency band on the wire or cable, and greatly reduced the signal power loading on repeater/amplifiers by removing the carrier power, thus minimising inter-modulation crosstalk between channels (see Chapter 10).

These advantages were also available on short-wave radio telephone systems, with the added benefit of reduced distortion due to multi-path radio transmission; many long-distance point-to-point commercial radio links were converted from double-sideband to single-sideband operation from the 1930s onwards (see Chapter 7).

Single-sideband, suppressed-carrier techniques became the dominant feature of virtually all wire, cable and microwave radio-relay frequency-multiplex transmission systems until the 1980s, when digital techniques began to be used (see Chapter 13), a remarkable tribute to the far-reaching influence of Carson's original mathematical analysis.

References

1 *A History of Engineering and Science in the Bell System: The Early Years, 1875–1925*, p. 277 et seq. (Bell Telephone Laboratories, 1975).
2 John Stone Stone, *The Practical Aspects of the Propagation of High-Frequency Electric Waves Along Wires* (Journal of the Franklin Institute, October 1912).
3 J.R. Carson, *Electric Circuit Theory and Operational Calculus* (McGraw-Hill, New York, 1926).
4 T.E. Shea, *Transmission Networks and Wave Filters* (Bell Telephone Laboratories, 1929).

Chapter 7

Pioneers of radio communication

7.1 The telegraph/telephone without wires

By the 1890s the theoretical predictions of Clerk Maxwell and the experimental demonstration by Heinrich Hertz of the existence of electromagnetic radio waves had begun to stimulate speculation that perhaps these 'Hertzian' waves might be used instead of wires to transmit telegraph and telephone signals over large distances. After all, light – which also consists of electromagnetic waves – had already been used in the semaphore system to transmit telegraph signals (see Chapter 2).

There is no doubt that the first major impetus towards the realisation of this dream was given by Guglielmo Marconi's experiments and demonstrations from 1894 onwards, but to which other workers in the UK, France and the USSR had made significant contributions.

Up to about 1910 the main emphasis was on the use of radio waves to transmit telegraph signals, but with the invention of de Forest's 'audion' triode thermionic valve in 1907, which made possible the generation and modulation of carrier waves (see Chapter 5), radio-telephony began to be studied. Despite the differences in the transmission media the technologies of wire and radio began to merge, including the use of carrier multiplex and single-sideband principles (see Chapter 6).

The successful evolution of radio communication over the century following 1890 has involved the detailed study of the propagation of radio waves over an ever-widening frequency spectrum ranging from about 30 kHz (10,000 metres wavelength) to at least 300 GHz (1 millimetre wavelength), and the creation of matching technologies to enable exploitation by a wide variety of radio-based communication services. These include point-to-point and mobile communication involving land, ship and air-borne stations, sound and television broadcasting (see Chapters 8 and 9), microwave radio-relay inter-city links (see Chapter

11), satellite communications (see Chapter 15) and 'cellular' mobile radio communication (see Chapter 21).

But to return to the pioneers who made it all possible.

7.2 The first radio-telegraph experiments

Guglielmo Marconi (1874–1937)

Guglielmo Marconi, the second son of a well-to-do Italian landowner, was born in 1874 at Bologna, Italy. For the most part the boy was brought up at the family's villa near Bologna, but received some of his early education from the age of five at a private school in Bedford, England. On his return to Italy he studied first at a school in Florence, and later completed his formal education at Livorno Technical High School where he studied physics.

Marconi displayed an original and inventive mind from an early age. When in his teens he came into contact with the renowned Professor Righi and studied his work on electromagnetic radiation. A critical moment in his life came when on holiday in the Italian Alps, he chanced upon a paper describing Hertz's radio experiments, and from this point was fired with the intention of finding a way of using Hertzian waves for communication. He was by no means a scientist looking for explanations of electromagnetic wave phenomena, but rather a practical

Guglielmo Marconi
At the time of his first experiments in England

A replica of Marconi's first transmitter which he used in his earliest experiments in Italy in 1895.

Figure 7.1   Marconi and a replica of his first transmitter

exploiter of radio waves for comunication at a distance. His early experiments were conducted at the family villa near Bologna, where his apparatus was similar to that used by other experimenters of his day, comprising a transmitter with an induction coil and a battery capable of generating high-voltage sparks across a short air gap in a Hertzian dipole radiator, and a receiver comprising a similar dipole coupled to a 'coherer'. The latter had been invented earlier by Professor Branly; it comprised a tube with two metal plugs and metal filings between the plugs. It had been christened 'coherer' by Sir Oliver Lodge because the metal filings tended to cohere in the presence of an electromagnetic wave train, thereby lowering the electrical resistance between the plugs and enabling current to flow in a battery circuit. The latter thus indicated the presence of the wave train, i.e. functioned as a 'detector'. To regain its sensitivity after the wave train had ceased the coherer required gentle tapping, provided by a trembler electric bell.

The transmission ranges achieved by Marconi at this time were no more than a hundred yards or so and he abandoned these experiments to study the electrical discharges generated by thunderstorms, using an elevated aerial – as the Russian physicist Popoff had also been doing at the time. Returning to his Hertzian wave experiments Marconi had the inspiration to use elevated aerials at the transmitter and receiver, each comprising a raised metal plate and a second plate on the ground, in place of the Hertzian dipoles. The improvement in range was magical – up to a mile and a half.

Encouraged by this success Marconi then set about making detailed improvements to the receiver, added a Morse key at the transmitter and a relay-operated Morse printer at the receiver, thus enabling telegraphic messages to be transmitted.

He was now in business as a communicator.

7.3 Early attempts at commercial exploitation

Marconi's first effort at commercial exploitation was to offer a demonstration to the Italian Government; it was a bitter blow when, after a delay, the Government declared itself not interested. After a family conference it was decided that Marconi, then a young man of 21 years, should go to England where the family had connections. There was another, and more powerful, reason for looking to England. He had early realised that one of the most valuable uses of his invention was by shipping – ships in distress had at that time no means of calling for assistance, either from the land or another ship. And England in the 1900s had the largest merchant fleet and most powerful navy in the world. His entry into England was hardly auspicious – an over-zealous Customs officer had broken his equipment. However, it was soon repaired and on 2 June 1896 Marconi applied for the world's first patent for wireless telegraphy (British Patent No. 12,039), granted on 2 March 1897. It is questionable whether the name 'invention' is strictly applicable to Marconi's system – he had in effect made use of ideas and devices originated by others but had combined and used them to achieve a commercially useful result.

With sound business instinct Marconi contacted Sir William Preece, Chief Engineer of the British Post Office and the British War Office, offering demonstrations. These first took place in 1896 between the Post Office Head-quarters building in St Martins le Grand and the Savings Bank building in Victoria Street, London, and later on Salisbury Plain where a range of one and three quarters of a mile was achieved. An observer from the Admiralty at the Salisbury Plain tests was Captain H. Jackson (later Sir Henry Jackson and First Sea Lord) who had already succeeded in sending wireless signals between two naval vessels. Sir William Preece lectured on Marconi's experiments to the British Association – his interest and support were however by no means dis-interested. The Post Office had been granted a Government monopoly in 1896 for telegraphic and telephonic communication in England and Preece, as Chief Engineer, saw wireless telegraphy in other hands as a possible usurper of that monopoly. He accordingly gave Marconi every opportunity to demonstrate his system, on the principle that to be forewarned is to be forearmed. Further experiments on Salisbury Plain followed in 1897 with kite- and balloon-supported aerials, and ranges of some four and a half miles were achieved. Meanwhile Professor A. Slaby of the Technical High School in Berlin, who was also working on wireless telegraphy, had taken a great interest in Marconi's experiments – an interest that later resulted in Marconi's greatest rival, the German Telefunken Company, formed by an alliance of the German General Electric and the Siemens and Halske Companies. [1] [2]

Realising the growing commercial pressures, Marconi arranged for the forma-tion in July 1897 of his own company to develop his system commercially; initially named 'The Wireless Telegraph and Signal Company Ltd', it later became Marconi's 'Wireless Telegraph Company Ltd' and finally 'The Marconi Company'.

Marconi continued his experiments; one established contact between an Ital-ian cruiser and the dockyard at Spezia over 11 miles and resulted in a decision by the Italian Navy to adopt Marconi's system. Another test, in co-operation with the British Post Office, established contact between Salisbury and Bath over a distance of 34 miles. Seeking to interest shipping organisations, he showed how a coastal station at Alum Bay on the Isle of Wight, equipped with a 120ft mast aerial, could maintain contact with pleasure steamers plying to Bournemouth and Swanage over a sea path of some 18.5 miles.

The Alum Bay demonstration was witnessed by Lord Kelvin in 1898; it must have given him considerable satisfaction and support for his electromagnetic wave propagation theoretical studies. It is recorded that Kelvin sent telegrams from the Alum Bay station via the mainland to Scotland, which he insisted on paying for, thereby both supporting Marconi's infant company and challenging the Post Office monopoly!

7.4 The development of 'tuning' – a great step forward

The spark transmitters then in use radiated a broad band of frequencies with the result that a receiver attempting to pick up weak signals from a distant transmitter might well be jammed by the stronger signals from a near-by transmitter. Such interference might well have disastrous consequences – for example to a ship in distress seeking help from a distant vessel. And as the numbers and power of transmitters grew, so the likelihood of interference increased. The problem was all the more acute for Marconi because rival systems in Germany and France were seeking a solution to this problem that would give them a competitive lead.

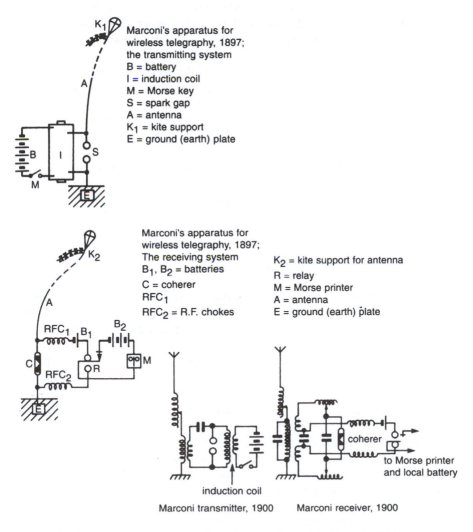

Marconi's apparatus for wireless telegraphy, 1897; the transmitting system
B = battery
I = induction coil
M = Morse key
S = spark gap
A = antenna
K_1 = kite support
E = ground (earth) plate

Marconi's apparatus for wireless telegraphy, 1897;
The receiving system
B_1, B_2 = batteries
C = coherer
RFC_1
RFC_2 = R.F. chokes

K_2 = kite support for antenna
R = relay
M = Morse printer
A = antenna
E = ground (earth) plate

induction coil

Marconi transmitter, 1900

Marconi receiver, 1900

coherer

to Morse printer and local battery

Figure 7.2   Marconi's transmitting and receiving equipment (1897 and 1900)

The solution to these problems was found by the principle of 'tuning', the use of resonant circuits to limit the spread of frequencies radiated by a transmitter and the band of frequencies to which a receiver would respond. An important lead was given by Sir Oliver Lodge's demonstration in 1889 of the principle of 'syntony' in which a receiving wire circle could be brought to resonance with a transmitting wire circle energised from a spark induction coil, by adjusting its effective length to correspond with that of the transmitter.

Marconi's further contribution – enshrined in his Patent No. 7777 of April 1900 – was to couple the transmitting aerial to the induction coil, and the receiving aerial to the coherer, via a 'high-frequency transformer', a tapped inductor in series with the aerial and a capacitor (Leyden jar) the capacitance of which could be varied to bring transmitter and receiver into resonance.

An important further development was that more than one transmitter could then be coupled to the same aerial, as could receivers to the receiving aerial.

The tuning concept in radio eventually developed similarly to the frequency-division multiplex pioneered by Campbell, Zobel and Carson for line and cable systems (see Chapter 6).

With the more important technical problems solved, the Wireless Telegraph and Signal Company began a major drive for commercial exploitation in Europe and America, initially with shipboard and coastal radio stations for government authorities such as the British Post Office, British and foreign navies and private shipping firms.

But a major challenge remained: could the Atlantic ocean be spanned by wireless telegraphy?

7.5 The Atlantic challenge for wireless telegraphy

Marconi's attempt to bridge the Atlantic by wireless telegraphy – when the prevailing scientific view deemed this to be impossible – was motivated by a desire to challenge both the submarine cable telegraph companies and the monopoly powers of the British Post Office.

He had no firm evidence to substantiate his belief that so great a distance could be spanned, beyond the fact some of his earlier experiments had shown that signals could be received some distance beyond the visible horizon. He reasoned that if a trans-Atlantic experiment was to stand any chance of success it would require the most powerful transmitter that could be built and the highest aerials that were practicable.

Marconi was fortunate in having the services and scientific advice of Dr J.A. Fleming, then a lecturer at University College, London, who was keenly interested in electromagnetic waves and also had experience in high-voltage alternating currents.

A site was chosen at Poldhu on the Lizard Peninsula, Cornwall in order to shorten the trans-Atlantic path (the first Post Office satellite Earth-station was also sited at Goonhilly, on the Lizard Peninsula and for the same reason – see Chapter 15). The spark transmitter was energised from a 25 kW alternator and keyed by short-circuiting part of the inductance in the alternator circuit. The transmitter power was thus far in advance of the battery-operated transmitters in use up to that time. The transmitting aerial system at Poldhu, and the receiving aerial at Cape Cod, USA, comprised as first built 200ft diameter circles of 20 masts, each 200ft high, designed by Marconi's assistant R.N. Vyvyan. Unfortunately both aerials were badly damaged by gales in September 1901. With much fortitude Marconi decided to continue with the work; his assistant Kemp built a temporary aerial system at Poldhu in the form of an inverted triangle of 50 copper wires some 160ft high. Marconi and Kemp then proceeded to St Johns, Newfoundland where a receiving system was set up in a hut on the appropriately named Signal Hill, not far from the landing of the first trans-Atlantic telegraph cable. Lacking a suitable mast, the 500ft-long wire receiving aerial was supported by a kite. The 'coherer' in the receiver was one designed for use by the Italian Navy; it comprised a globule of mercury between carbon electrodes, and probably functioned more as a rectifier than the usual iron filing coherer. A telephone-type earpiece was used to listen to the received signals, providing greater sensitivity than was given by the usual relay and Morse printer. [3]

There is uncertainty about the operating wavelength of the system, since wavemeters were not available at that date; one estimate put it at 366 metres, but it may have been much longer.

Note: *The first faint signals from Poldhu, in the form of the three dots of the Morse Code for the letter 'S', were received on 12th December, 1901.*

There was at first considerable scepticism about Marconi's claim to have bridged the Atlantic by wireless telegraphy – partly from the mathematical physicists who claimed that, even allowing for diffraction around the Earth, it was impossible and partly because, in the absence of a record from a Morse printer, the claim rested wholly on the unsubstantiated words of Marconi and his assistant Kemp. And uncertainty was compounded when it was found that the results were not always repeatable, there being a marked difference between the signal strength when the trans-Atlantic path was in darkness and when it was in daylight. It later became clear that an electrified layer in the upper atmosphere, predicted by Heaviside in England and Kennelly in the USA, was playing a significant role in guiding the radiated waves across the Atlantic.

Some light was thrown on the wave propagation problem by ship-borne receiving tests across the Atlantic on transmissions from Poldhu carried out by Marconi in 1902. These revealed that whereas in daylight the range was limited to some 700 miles, this increased to over 2,000 miles at night. This result indicated that there was at least a good prospect of maintaining contact with shipping in mid-Atlantic both by day and by night.

7.6 The further development of long-wave wireless telegraphy

From 1902 to the 1920s many important developments took place in long-wave wireless telegraphy, and have been described by R.N. Vyvyan, Marconi's Chief Engineer. These included greatly increased transmitter power and the use of continuous waves in place of spark transmissions, at first derived from Alexanderson high-frequency alternators and later from high-power water-cooled thermionic valves. Receiver sensitivity was gradually improved, at first by using a 'magnetic' detector comprising a moving magnetic tape in place of the coherer, then a point-contact 'crystal' rectifier, and lastly the Fleming diode or de Forest triode valve (Chapter 5). [3]

Following objections by the Anglo-American Cable Company to Marconi's operations at Newfoundland he built, with the agreement of the Canadian Government, a high-power transmitting station at Glace Bay, Canada. He also moved his operations from Poldhu to Clifden on the West Coast of Ireland. However, following a disastrous fire in 1909 and troubles in Ireland, this was replaced by a high-power transmitting station at Carnavon, Wales that commenced public wireless telegraph service in 1914.

European Governments, notably in England, Germany and France, had realised the strategic importance of long-distance wireless telegraphy in time of war and the vulnerability of the undersea cable telegraph system – then the only

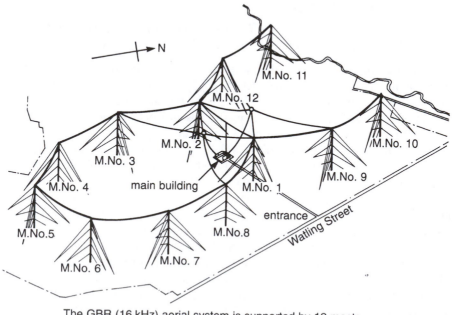

The GBR (16 kHz) aerial system is supported by 12 masts,
each 820 ft high. It occupies a site $1^1/_4$ miles long and $^3/_4$ mile wide.

Figure 7.3   Post Office (British Telecom) long-wave radio transmitting station at Rugby

means of inter-continental communication. High-power, long-wave wireless telegraphy stations were built in several countries. In the UK the Post Office Engineering Department designed and built the long-wave telegraphy transmitter GBR at Rugby – operating on a frequency of 16 kHz, some 20,000 metres wavelength, with a vast aerial array suspended from 800ft high masts and extending over two miles. It is one of the most powerful radio stations in the world, and provides broadcast telegraphy for the press and news services and time signals with a world-wide range extending to the antipodes. This was an interesting application of a principle that Marconi had discovered – the range of long-wave signals guided between the Earth and the Heaviside/Kennelly layer increases as the wave-length is increased.

By 1926 a second high-power transmitter at Rugby, GBY, operating on 50 kHz, 6,000 metres wavelength, built by the British Post Office in co-operation with the ATT Co.mpany, provided a single both-way telephone circuit between the UK and USA. It was an early application of the single-sideband technique originated by J.R. Carson and the wave filters due to Campbell and Zobel of Western Electric (Chapter 6).

In the UK plans for an 'Imperial Wireless Chain' had been made in 1913 with the aim of linking the UK with Canada, Egypt, India, the Far East and Australia via a chain of high-power long-wave transmitters, the first of which was to be located at Leafield, Oxfordshire where the Post Office already had a long-wave transmitter handling its ship-shore wireless telegraphy traffic. However, before the Imperial Wireless Chain could be completed,a new technique of long-distance radio communication became possible – short-wave beamed radio communication via ionospheric reflection – that offered significant advantages compared with long-waves. One of Marconi's protegés, C.S. Franklin, played a major role in ushering in the short-wave era. [3]

7.7 The era of beamed short-wave radio communication

C.S. Franklin (1879–1964)
The limitations of the long-wave radio system were all too apparent – for long distance transmission the aerial systems were massive, the transmitter powers needed were large and the limited space available in the frequency spectrum precluded more than a few telegraph services, with almost no room for telephone services. Moreover, with long-wavelengths it was impracticable to concentrate transmitted wave energy into a beam since the aerial dimensions required would have been excessive. Hertz's experiments in 1886 had already demonstrated that short-wavelength radio waves could be concentrated into a beam by parabolic mirrors (see Chapter 2), and this no doubt led Marconi to arrange for C.S. Franklin to carry out an intensive study of the propagation characteristics of short-wavelength waves and their possibilities for long-distance telegraph and telephone communication.

By 1919, thermionic valves capable of generating a few hundred watts of

C. S. Franklin

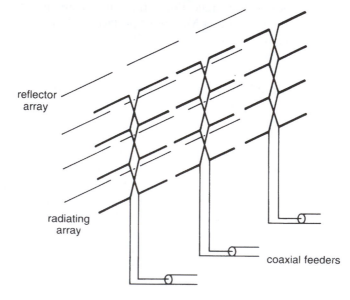

Figure 7.4   Franklin and short-wave beam array

continuous wave power at wavelengths down to about 15 metres had been developed and were used initially for overland tests in England that achieved ranges of up to about 100 miles with the aid of parabolic reflector aerials. It was then decided to carry out a larger scale trial, using some 10 kW of power on a wavelength of 100 metres and a parabolic aerial mounted 325ft high on a mast at Poldhu. The receiver, which did not use a reflector at the aerial, was mounted on Marconi's yacht *Elettra*. By sailing down into the South Atlantic it was shown that ranges of 1,250 nautical miles by day and at least 2,230 miles by night were achievable. Other tests in 1924 on 100 metres wavelength with a power of about 20 kW at Poldhu reported good night-time reception in New York, and in May of that year speech was for the first time successfully transmitted from England to Australia.

It must in fairness be pointed out that radio amateurs, who had been given wavelengths below 200 metres because they were thought to be useless for commercial purposes, also began to demonstrate long-distance contacts, often with powers of only a few watts and aerials of limited directivity.

Instead of the somewhat cumbersome parabolic reflectors used in the early experiments Franklin devised a highly efficient 'curtain array' directional aerial system, based on half-wavelength co-phased wire elements arranged in vertical or horizontal groups, and linked to the transmitter/receiver by non-radiating low-loss coaxial feeders. This mechanically convenient, efficient and economic solution to aerial design contributed substantially to the development of short-wave radio communication.

These results gave Marconi and his team confidence to propose to the British Government an 'Imperial Beam System' based on beamed short waves to replace the earlier planned long-wave high-power system – offering lower costs, better performance and greater flexibility. The Government, to avoid a conflict of interest between cable and wireless companies and to ensure more effective utilisation of both media, brought both together in a new company, Imperial and International Communications Ltd, later re-named Cable and Wireless Ltd with the obligation to provide world-wide telegraphic communication. In the UK short-wave transmitting stations equipped with Franklin-type aerial arrays were built at Ongar and Dorchester, and receiving stations at Brentwood and Somerton. These stations worked to similarly equipped overseas stations, many supplied by, or built under licence from, the Marconi Company.

7.8 Development of beamed short-wave radio-communication by engineers of the ATT Co.

Recognising the limitations of a single-channel trans-Atlantic radio-telephone link using long-waves (60 kHz), and noting the encouraging results from Marconi's short-wave tests from Poldhu, England, engineers of the ATT Co., Western Electric and Bell Laboratories in the USA began to explore the possibilities of short-wave radio in the 1920s. An experimental transmitting station

was established at Deal, New Jersey and the development of high-power transmitters put in hand, so successfully that by 1930 a 60 kW output stage using
water-cooled valves and driven by linear lower-power radio-frequency amplifiers had been achieved.

Short-wave receiving stations were set up at Holmdel and Netcong, New
Jersey. [4] Prominent amongst the engineers who made important contributions
to this work were: A.A. Oswald; M.J. Kelly; W. Wilson; R.A. Heising; J.C.
Schelleng; H.T. Friis; E.J. Sterba; E. Bruce.

Considerable effort was put into the development of short-wave arrays in view
of the advantages of the gain in effective radiated power and the reduction of
interference offered by the array directional properties. One design – the 'Sterba
Array' – was similar to the Franklin vertical curtain of half- and quarter-wave
elements. Another – the 'Bruce Array' – at Netcong, New Jersey used a horizontal group of vertical half-wave elements with a similar reflecting group a
quarter-wavelength away. [5]

A steerable version of the Bruce array was used by K.G. Jansky in his
discovery of inter-stellar radio-frequency noise radiation in the Milky Way. [6]

E. Bruce, with H.T. Friis, was responsible for the most widely used of all
short-wave arrays – the 'Horizontal Rhombic Aerial', which largely displaced
the large curtain arrays for commercial point-to-point radio services. The horizontal rhombic aerial had the substantial advantages of simplicity and low cost
– it could be supported on poles some 80 ft high, in contrast to the expensive
structures needed to support the vertical curtain arrays. Moreover, it was effective over a frequency range of about 4 to 1, e.g. from 5 to 20 MHz, as compared
with the relatively narrow bands offered by resonant element systems; effective
power gains of some 14 decibels could be realised. [7]

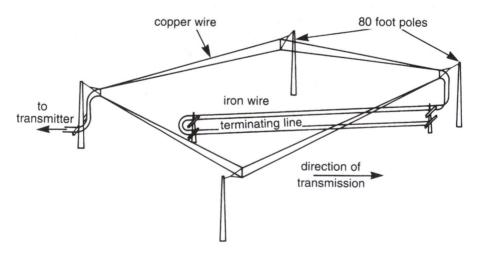

Figure 7.5   The short-wave rhombic aerial (ATT Co.)

(Shown as used for transmission, also used for reception with resistor termination)

From the mid-1920s to the 1930s exhaustive radio-wave propagation tests were made, mainly on the trans-Atlantic path to England where a receiving station had been established by ATT Co. at Southgate, near London, and in co-operation with the British Post Office at their receiving station near Baldock, Hertfordshire. These tests aimed at providing data to determine the optimum frequencies for use at various times of day and seasons of the year, taking into account the varying influence of solar sun-spot activity on the electrified E and F layers of the ionosphere, which formed the Heaviside/Kennelly layer referred to earlier.

Experiments in which multi-tones spaced at about 100 Hz were transmitted in the voice band clearly revealed the existence of selective fading due to multi-path transmission via the various layers and multiple-hop paths, and the corresponding delay spread of the received signal components by up to several milliseconds. The speech distortion due to multi-path transmission was one of the limitations inherent in short-wave transmission via the ionosphere. [8]

Developments in Europe, notably in the British Post Office, had closely paralleled those in the USA. The ATT Co. built a transmitting station for commercial operation at Lawrenceville, New Jersey, while the British Post Office established one at Rugby.

In September 1929 a demonstration was given, in conjunction with the British Post Office in which Bell Laboratories officials in New York talked with Australia via London – an historic occasion which marked the beginning of the short-wave point-to-point radio communication era.

The first commercial radio-telephony services were operated on a double-sideband amplitude-modulation basis; however, it had been realised for some time that single-sideband working offered a potential 9 decibels improvement of signal-to-noise ratio (6 db from suppression of the carrier wave, and 3 db from the reduction of bandwidth by removal of one sideband), equivalent to an eight-fold transmitted power increase. An additional advantage was the removal of non-linear distortion that occurs in double-sideband operation when the carrier fades relative to the sidebands. The problems of frequency stabilisation at the receiver to enable the re-inserted carrier frequency to be corrected within a few cycles per second were eventually solved and single-sideband working began to be used generally from the mid-1930s. Further developments enabled four telephone channels to be provided two on either side of a partially suppressed carrier, described as 'independent sideband' operation.

The last major technological development in the short-wave era was the 'MUSA' – the 'Multiple Unit Steerable Antenna' for short-wave reception – which had the aim of further improving the quality and reliability of short-wave point-to-point radio communication by separating the rays due to multi-path propagation and then combining the demodulated voice signals after delay correction.

This brilliant concept was due to the pioneering group of Bell Laboratories scientists and engineers at Holmdel, New Jersey and has been described by H.T. Friis and C.B. Feldman. A matching MUSA was designed and built by the

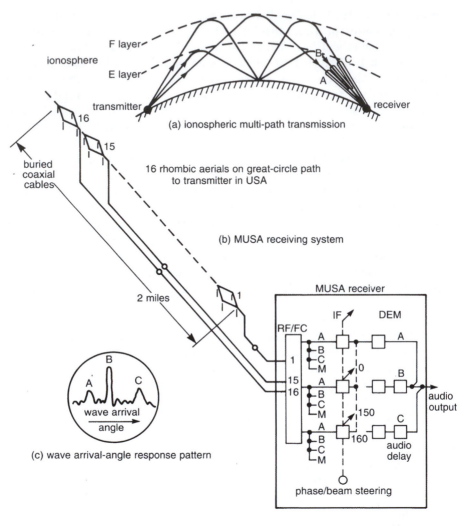

Figure 7.6   *The multiple-unit short-wave steerable array MUSA (1940–1960)*

British Post Office for the trans-Atlantic link. The story of the Bell System contributions to the short-wave era ends on a somewhat wistful note:

A sentimental reunion ceremony was held for the many engineers who contributed to the development of the short-wave art on the closing down of the Lawrenceville (transmitting) station in February 1976. Most of the people had already retired, but history would have been kinder to them if it had taken a bit longer for the short-wave radio-telephony art to become obsolete. [4] [9]

Nevertheless, even today the short-wave radio spectrum is crowded with commercial, broadcasting and amateur transmissions – a lasting tribute to the pioneers whose work made it possible.

7.9 Development of beamed short-wave radio-communication by engineers of the British Post Office

The Government had placed on the British Post Office the obligation to develop long-distance overseas radio-telephony services to and from the UK, with Cable and Wireless Ltd handling the overseas telegraphy services. The equipment development for radio telephony became the responsibility of the Engineering Department of the British Post Office and this development was carried out under the direction of its Engineers-in-Chief, notably G. Lee, C. Angwin, A.J. Gill and A.H. Mumford, all of whom were knighted for their services. In addition to the long-wave transmitters GBR and GBY referred to above, they were responsible for building short-wave transmitting systems at Rugby and receiving systems at Baldock. Their managerial support for the innovations reported here were undoubtedly an important contribution to a successful outcome.

The pioneering work of the Post Office engineers in the development of beamed short-wave radio-telephony systems from about 1920 to 1950, included the design and construction of:

- aerial arrays;
- high-precision quartz-crystal drives for transmitter and receiver frequency control;
- high-power transmitters;
- super-heterodyne receivers;
- single-sideband and independent-sideband (4-channel) transmitter drives and receivers.

The general development of radio-communication in the British Post Office has been described elsewhere. In much of this work there was close liaison with engineers of the ATT Co.mpany and the Bell Telephone Laboratories. Some developments such as single- and independent-sideband working were first tried out on the trans-Atlantic route and later extended to other overseas routes. [10] [11]

As noted above, the British Post Office in co-operation with the ATT Co. built a multiple-unit steerable antenna (MUSA) receiving system for the trans-Atlantic link, described by A.J. Gill in his address as Chairman of the Wireless Section of the Institution of Electrical Engineers, London in 1939. [12]

The MUSA represented the ultimate development in short-wave receiving systems – it comprised 16 horizontal rhombic aerials arranged on the Great Circle path pointing towards the distant transmitter, the radio-frequency outputs of which were passed over low-loss coaxial feeders to the receiving installation. There, after conversion to an intermediate frequency, the signals were adjusted to combine in phase. By so doing the extremely sharp spatial directivity was achieved in the vertical plane, enabling individual down-coming rays reflected from the E and F layers of the ionosphere to be selected and the demodulated voice signals then to be combined. The overall effect was to improve the received signal-to-noise power ratio by up to 30 times, together with an improvement of voice quality by minimising distortion due to radio wave path time-delay differences.

The Post Office MUSA was located on the Cooling Marshes in Kent, whilst the ATT Co. built their system at Manahawkin, New Jersey. The British system was unique in that it used an electrical phase-shifting system based on an artificial long line with 16 tappings, whereas the American system used a mechanical system with a 16-step gear box. The British phase-shifting system was in some ways a pre-cursor of the phase-shifting arrays later developed for radar.

Although the hoped for improvement in trans-Atlantic communication reliabilty was not always achieved – as for example when the ionosphere was badly disturbed due to solar activity – the MUSAs nevertheless gave valuable service from 1940 to 1960. In the absence of submarine telephone cables this service was especially significant during the war years when direct communication between the US and UK Governments was needed. For example, it enabled Prime Minister Churchill and President Roosevelt to have direct and immediate telephone communication – using high-grade speech-privacy equipment – especially important when speed of decision was critical.

7.10 The role of the short-wave radio communication services

The long-distance short-wave radio telephone and telegraph services undoubtedly played a vital role in meeting the need for ever-expanding world-wide communications from the 1920s to the 1960s. For this the named, and the many un-named, pioneers who made it possible deserve our thanks and appreciation – it is easy to forget that sometimes, as in Marconi's early experiments, they were working in a field where scientific knowledge was often lacking and progress had to be made by engineering intuition.

The short-wave services were, however, subject to interruption and loss of voice quality at times of ionospheric storms, and the scope in the radio-frequency spectrum for continued growth of services was itself limited. By the 1960s multi-channel telephony submarine cables were coming into service and satellite communication was soon to follow.

The era of long-distance short-wave beamed point-to-point radio communication is coming to a gradual close; nevertheless, for certain applications, such as world-service broadcasting, maritime and defence services short-wave radio has a continuing role.

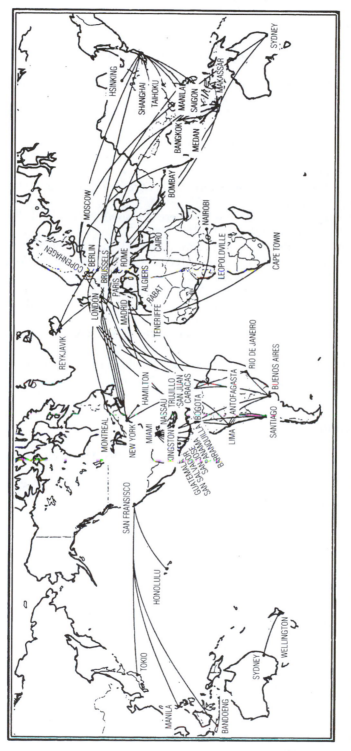

Figure 7.7 International radio telephone circuits of the world (1937)

The first circuit, New York–London, opened on long-waves in 1927

References

1 W.J. Baker, *A History of the Marconi Company* (Methuen and Co. Ltd., 1970).
2 E.C. Baker, *Sir William Preece, FRS, Victorian Engineer Extraordinary*, 1976.
3 R.N. Vyvyan, *Wireless Over Thirty Years* (George Routledge and Sons Ltd., 1933).
4 *A History of Science and Engineering in the Bell System: Transmission Technology (1925–1975)*, edited by E.F. O'Neill (Bell Laboratories, 1985). Chapter 2, Overseas Radio.
5 E. Bruce, 'Developments in Short-Wave Directive Antennas', (*Bell Syst. Tech. Jl.*, 10 October 1931).
6 K.G. Jansky, '*Electrical Disturbances Apparently of Extra-Terrestrial Origin*' (*Proc. IRE*, October 1933).
7 'Horizontal Rhombic Antennas', (*Proc. IRE*, 23, January 1935).
8 R.K. Potter, 'Transmission Characteristics of a Short-Wave Telephone Circuit', (*Proc. IRE*, 18, April 1930).
9 A.A. Oswald, 'The Manahawkin MUSA' (*Bell Lab. Rec.* 18, January 1940).
10 'The Development of Radio-Communication in the Post Office', (*P.O. Electrical Engineers Jl.* 49 Pt.3, October 1956).
11 W.J. Bray *et al.*, 'Single-Sideband Operation of Short-Wave Radio Links', (*P.O. Electrical Engineers Jl.*, 45, October 1952–October 1953).
12 A.J. Gill, Chairman's Address, Wireless Section, Institution of Electrical Engineers, London, (*Jl.IEE*, 84 No 506, February 1939).

Further Reading

G. Bussey, *Wireless: The Crucial Decade 1924–34*, (IEE History of Technology Series No. 13).
F. Rowlands and J.P. Wilson, *Oliver Lodge and the Invention of Radio* (PD Publications, London, 1994).

Chapter 8

Pioneers of sound radio broadcasting

8.1 The beginning of sound radio broadcasting in the USA

Once the problems of generating a continuous carrier wave and modulating it by speech had been solved and applied in point-to-point and ship-to-shore services it was perhaps inevitable that enterprising amateurs and entrepreneurs should seek to explore the possibilities of the new technology for communication with a wider audience.

The first broadcasting stations appeared in the USA where the absence of strict Government regulation made it possible for anyone to set up and operate a radio station, requiring little more than the registration of the wavelength to be used – at first around 360 and 400 metres. Lee de Forest – whose name is associated with the invention of the triode valve (Chapter 5) – was one of the first to set up a low-power amateur station in 1907.

In 1919 the Radio Corporation of America had come into being having acquired substantial patent rights from the British Marconi Company, the de Forest patents on the triode valve and Westinghouse patents of Armstrong and Fessenden on the principles of heterodyne reception (use of a non-linear device to mix a carrier and local oscillation to generate an intermediate frequency) and regeneration (feedback to improve reception sensitivity).

The RCA was thus in a strong position commercially to exploit sound radio broadcasting – but its success in so doing owed much to an organizing genius, David Sarnoff.

David Sarnoff (1891–1971)
David Sarnoff, born in 1891, joined the American Marconi Company as an office-boy in 1906 but soon displayed a keen interest in wireless telegraphy, carrying out experiments in his spare time. He became a ship's wireless operator and gave valuable services at the time of the Titanic disaster. By 1916 he had risen to the position of Contracts Manager in RCA and made the then remarkable proposition to the company that radio transmitting stations should be built

for the purpose of broadcasting speech and music, and a 'radio music box' be manufactured for sale to the general public.

The proposal was revolutionary and had breathtaking implications – the very idea of radio for 'entertainment' was new but if the public found it acceptable the mass market for receivers could be enormous and commercially very profitable. Newspaper owners, manufacturers and retail store owners were quick to realise the potential of broadcasting for advertising their wares, and by 1920 a boom in broadcasting had started in the USA.

One of the most well-known broadcasters was the high-power station KDKA in Pittsburg, Pennsylvania built by Westinghouse, partly to stimulate its sales of radio receiving equipment.

By 1924 there were more than 1,000 stations in operation in the USA, and since the few wavelengths then available had often to be shared by several stations there was a great deal of mutual interference. To remedy this chaotic situation, and to impose a degree of control over the utilisation of the frequency spectrum, the US Federal Communications Commission was eventually formed.

8.2 The beginning of sound radio broadcasting in Britain

In Britain sound radio broadcasting got off to a slower, but better controlled, start. The British Post Office was responsible to Parliament for the regulation of wireless telegraphy, and this was held to include radio broadcasting. At first the Post Office licensed only low-power transmitters for experimental purposes, but by 1922 had authorised the Marconi Company to carry out limited broadcasting from an experimental station at Writtle, near Chelmsford, with a maximum power of 250 watts. [1]

The Writtle station operated on a wavelength of 400 metres with a four-wire aerial 250 ft long and 110 ft high, the transmitter being adapted from a standard radio-telephone transmitter of the time. The station call sign was 2 MT – pronounced 'Two Emma Toc'. The licence for the experimental station permitted only a meagre half-hour of broadcasting a week, but even this did not deter the small group of Marconi engineers, led by Capt. P.P. Eckersley, who built and operated the station.

Capt. P.P. Eckersley (1892–1963)
P.P. Eckersley and T.L. Eckersley (see below) were grandsons of the great Victorian scientist Professor Thomas Henry Huxley, and there is little doubt that some of the genius of their illustrious forebear 'rubbed off' on the Eckersleys.

The first 'entertainment broadcast' was made from the wooden hut at Writtle on 14 February 1922, to the joy of the many amateurs who had pressed for a radio broadcasting service. And P.P. Eckersley proved to be not only a competent engineer but a first-class entertainer, gifted with spontaneous humour that brought enthusiastic responses from the listening audience.

These early, and necessarily brief, programmes included not only a

Capt. P. P. Eckersley

An early wireless transmitter, 2 MT, Writtle, 1921

Figure 8.1   Eckersley and the first experimental broadcasting transmitter at Writtle, 1921

The Writtle Pioneers
Left to right, Standing: B. N. MacLarty, H. L. Kirke, R. T. B. Wynn, H. J. Russell;
Seated: F. W. Bubb, Noel Ashbridge, P. P. Eckersley, E. H. Trump, Miss E. M. Beeson

Figure 8.2   The Writtle pioneers of the birth of broadcasting in the UK

rudimentary *Children's Hour* – of which P.P. Eckersley was 'Uncle' – but also the first radio play (*Cyrano de Bergerac*), performed via a hand-held carbon microphone passed from performer to performer. And it was from Writtle that Dame Nellie Melba made an historic first operatic broadcast.

And so, in spite of a lack of official enthusiasm, British sound radio broadcasting was born – mainly through the efforts of a few pioneers with vision and drive. To quote from W.J. Baker [2]:

On 17 January 1923, the Marconi station 'Two Emma-Toc' at Writtle had made its final bow to a regretful circle of enthusiasts. It closed down with full honours, its mission accomplished; it had created an enthusiasm for broadcasting that was destined to make it into the greatest medium for entertainment and instruction that the world had ever seen. In its short and hilarious career it had laid the foundations of the age of broadcasting.

Meanwhile the Post Office had authorised the Marconi Company to set up an experimental station, call sign 2LO, at Marconi House in London with a maximum power of 1.5 kW and a wavelength of 360 metres. Transmission times were at first limited to one hour daily but the programmes began to draw sizeable audiences by including music concerts and outside broadcasts of special events.

Capt. H. J. Round	T. L. Eckersley, FRS
Circuit and device inventor	Outstanding radio wave propagation theorist

Figure 8.3   Round and Eckersley – radio-communication pioneers of the Marconi Company

8.3 The British Broadcasting Company is created

As was to be expected, the granting of an experimental radio broadcasting licence to the Marconi Company brought many applications from other UK manufacturing companies for similar facilities. To avoid the possibility of broadcasting chaos – the signs of which were already apparent in the USA – the Postmaster General invited all interested parties to form a consortium to create a single broadcasting authority for the UK.

The outcome was the British Broadcasting Company Ltd, formed from the Marconi's Wireless Telegraph Co. and five other manufacturers, with revenue derived from a tariff on all broadcast receiving equipment sold by them and half the 'ten shilling' licence fee paid by all listeners.

P.P. Eckersley joined the BBC in 1923, becoming its Chief Engineer until he left the company in 1929. His major achievement was the 'Regional Broadcasting' system in which twin high-power transmitters at each site radiated two programmes on different wavelengths.

The growth of new stations was phenomenal – stations in Birmingham, Manchester, Newcastle, Cardiff, Glasgow, Aberdeen, Bournemouth and Sheffield followed in quick succession.

P.P. Eckersley had a visionary quality that is the stamp of greatness, a charmimg and powerful personality and a remarkable ability to deflate over-pompous individuals. He did more than anyone to put the technical foundations of broadcasting in the UK on a sound footing. His book *The Power Behind the Microphone*, [3] gives a vivid account of the problems – organisational and technical – that he fought to overcome. However, this brief account of the newly formed broadcasting company would be incomplete without mentioning another BBC engineer whose innovative abilities played an important role in its development, Captain H.J. Round.

Captain H.J. Round (1881–1966)
H.J. Round studied at the Royal College of Science, London where he gained a first-class honours degree. He worked initially as an assistant to Marconi on direction-finding and long-wave transmitter design. Before the First World War he became involved in improvements to thermionic valves – including the use of oxide-coated filaments and indirectly-heated cathodes, taking out numerous patents in 1913–1914. During the war years he became responsible for setting up direction-finding networks on the Western Front and in the UK – the latter detected the movement of the German Fleet from its base in Wilhelmshaven that signalled to the British Admiralty that the Battle of Jutland was imminent. As medium-wave broadcasting expanded territorially the need for higher transmitting power to extend the coverage of individual stations became apparent, and in this area Captain H.J. Round made important contributions. He also furthered the development of microphones and receivers, including the 'reflex' circuit which enabled a single valve to operate simultaneously as a high-frequency and an audio-frequency amplifier.

8.4 The reflecting layers of the ionosphere are revealed

As information concerning the performance of medium-wave broadcasting transmitters began to build up it became apparent that after nightfall the range over which signals could be received was often greatly extended, but in this extended range there was both fading and distortion. Evidently a 'sky-wave' was appearing at night, possibly due to reflection from an elevated atmospheric layer or layers. This phenomenon also caused interference when transmitters within an extensive geographical area such as Europe shared the same wavelength. Clearly more information as to the nature and properties of the reflecting layers was needed, partly to determine the extent to which wavelengths could be shared but also to enable transmitting aerials to be designed for minimum sky-wave radiation. Dr E. Appleton of King's College, London carried out a significant experiment in the 1920s in which the frequency of the BBC transmitter 6BM at Bournemouth was gradually shifted over a small range and the interference fringes between the ground-wave and reflected sky-wave were observed at a suitably placed receiving station. The interference fringes revealed a reflecting layer at a height of about 100 km, named by Appleton as the 'E-layer' (not E for Edgar, Appleton disclaimed, but E for 'electron'!).

In the USA Breit and Tuve of the National Bureau of Standards Laboratory transmitted pulsed signals vertically and demonstrated the multiple layer structure of the ionosphere. It became clear that there were other layers, named F1 and F2, above the E layer, and an absorbing 'D' layer below it, the behaviour of which varied daily, anually and over an 11-year cycle linked with sun-spot activity.

In view of its great value for the planning and operation of radio broadcasting and point-to-point communication systems, the scientific study of radio-wave propagation and the ionosphere became an important responsibility of the Radio Research Station of the National Physical Laboratory in the UK, and the National Bureau of Standards in the USA.

Important contributions to this study were made in the 1920s and 1930s by T.L. Eckersley of the Marconi Company.

T.L. Eckersley, FRS (1887–1959)

T.L. Eckersley acquired a BSc at University College, London and did research at the Cavendish Laboratory, Cambridge and the National Physical Laboratory.

During the First World War he worked with the Royal Engineers, carrying out theoretical and experimental radio studies on the 'night effect' of sky-waves and the 'refraction' radio waves suffer at coasts and which affects radio direction-finding. He joined the Marconi Company in 1919 and commenced the intensive study of radio-wave propagation that became his life's work.

With the advent of the Marconi-Franklin beam system of short-wave communication he turned his attention to the long-distance propagation of high-frequency (short-wave) electromagnetic waves.

During 1924 to 1934 he directed a research team that carried out many pioneering theoretical and experimental wave propagation studies. This work resulted in a number of papers for the Institution of Electrical Engineers that laid down a basis for the prediction of the performance of high-frequency radio services that was brilliantly confirmed in practice.

T.L. Eckersley's genius led him to apply the 'phase-integral' method of analysis, familiar in quantum mechanics, to the magneto-ionic theory of ionospheric propagation and the diffraction of radio waves around the curved Earth. The work was later extended to include tropospheric-scatter propagation at very high frequencies.

In 1938 he was made a Fellow of the Royal Society for these outstanding contributions to the study of radio-wave propagation, and he became a Fellow of the American Institute of Radio Engineers in 1946. The citation for the latter reads:

Both his approach to the problem from the standpoint of communications and his invention of mathematical tools useful in the computation of radiated fields are achievements of lasting value, acclaimed by the whole radio world and of which he may be justly proud.

8.5 Broadcasting on very high frequencies

Following the end of the Second World War it became apparent that sound radio broadcasting would need to expand beyond the numbers of transmitting stations that could be accommodated in the long- and medium-wave bands. Attention was then directed towards the Very High Frequency (VHF) band 87.5–100 MHz that had been allocated to broadcasting by international agreement, and which offered much greater scope for growth. It also enabled the quality of reception to be improved by increasing the band of audio frequencies that could be transmitted, compared with long- and medium-waves, to 20Hz–15 kHz needed for high-fidelity music transmission. And interference from distant transmitters could be greatly reduced, partly because of the more stable radio-wave propagation characteristics of VHF, and partly because directional aerials were practicable. However, since the range available did not extend greatly beyond the visible horizon, more transmitters albeit of lower power were needed to provide a service over a given area. Much of the technology developed during the war years for radar and fighting services radio communication was available or could be adapted for broadcasting.

But an important question remained – what system of modulation should be used for public service VHF broadcasting in the UK, the well-tried amplitude modulation used on long- and medium-waves, or frequency-modulation which by 1945 already had considerable application in the USA?

8.6 Sound broadcasting acquires enhanced quality

8.6.1 The invention of frequency modulation

Major Edwin Armstrong
Armstrong in the USA had examined the use of frequency modulation of a radio carrier wave in the early days of long-wave radio-telegraphy in an attempt to eliminate static interference. His experiments were not successful and a mathematical analysis by J.R. Carson of Bell suggested that there was no band-width reduction or signal-to-noise ratio improvement to be gained by the use of frequency modulation (FM) compared with amplitude modulation (AM).

However, Armstrong pursued his study of FM and, by excellent engineering intuition, introduced the all-important concept of an 'amplitude limiter' in FM reception (this had been omitted in Carson's mathematical model). The effect was almost magical – by arranging that the peak-to-peak frequency deviation of the FM carrier wave was several times the highest audio frequency to be transmitted, most of the noise background at the FM receiver output fell outside the wanted audio band and a substantial improvement in signal-to-noise ratio was achieved. An additional bonus was a 'capture effect' which caused a weaker interfering carrier to be virtually eliminated.

Armstrong had a long struggle to perfect his new system and to convince American manufacturers and broadcasters of the value of his invention, but by 1945 the widespread use of FM in the USA was proof of its worth.

In Britain interest in FM was no doubt stimulated by a long-standing friendship between Major Armstrong and Captain Round of the Marconi Company. After the end of the war the BBC Research Department, headed by H.L. Kirke, the Marconi Company and the British Post Office began an extensive series of tests to assess the advantages of FM and to optimise the technical characteristics to be used in the public broadcasting system.

By 1950 the first BBC 25 kW FM transmitter, built by the Marconi Company, commenced operation at Wrotham Hill, Kent and a new high-quality public broadcasting service in the UK began. [1] [2] [3]

Frequency modulation also had an important role in the VHF mobile radio services used by ambulance, police and other public authorities and by private users that began to be developed in the post-war years.

8.7 Sound broadcasting acquires two-dimensional reality

8.7.1 Stereophonic broadcasting and recording

Alan Blumlein (1903–1942)

Sound quality and realism in radio broadcasting and record reproduction have been greatly enhanced by stereophony – in which the reproduced sound appears to originate from sources distributed in space as were the original sources, rather from a single point in space. This is in addition to the wider audio-frequency band and lower noise levels available in FM broadcasting.

Whilst the research organisations of broadcasting authorities have made useful contributions in this field, there is no doubt that a major and initial impetus came from a British inventive genius – Alan Blumlein.

Alan Blumlein was born in 1903, educated at Highgate School and graduated at City and Guilds Engineering College, London – his first-class honours degree being in heavy-current electrical engineering. In 1924 he joined Standard Telephones and Cables Ltd, where his early work was concerned with the improvement of transmission on long-distance telephone lines. In 1929 he joined the Columbia Record Company, which later merged with His Master's Voice record company to form EMI (Electric and Musical Industries Ltd). American record companies had made considerable improvements in sound recording techniques and Blumlein was asked by the EMI Managing Director, Isaac Schoenberg, to study ways of circumventing the American patents – this Blumlein did and produced even better techniques for driving the record-cutting stylus.

In the 1930s the Bell Telephone Laboratories in the USA had been studying methods for recreating a more realistic sound field using multiple loud-speakers. Recognising that a simpler approach was commercially desirable Blumlein concentrated on 'deceiving the ear' by a pair of loud-speakers energised from two microphones, each having a directional pattern, and with the pattern axes arranged at 90 degrees – the 'crossed figure-of-eight'. Blumlein's 1931 patent covers this and other microphone configurations, but even more importantly, means for recording pairs of stereo signals in a single record groove or a single film track.

However, it was many years before the record companies and the film industry took stereophony seriously and it is doubtful whether Alan Blumlein ever received an adequate reward for his inventions.

Blumlein's vital contributions to the development of television are described in Chapter 9; in addition to these he carried out important work on air-borne radar ground surveillance techniques in the war years. Sadly he died in an air crash in 1942 while testing radar equipment. His colleagues described him as at times a difficult man but with a remarkable ability to cross boundaries between scientific philosophies and between groups of people.

A detailed account and appreciation of A.D. Blumlein's work, and an illuminating pen-picture of his personal qualities, has been given by Professor R.W. Burns in. [4] It assesses his contributions:

A.D. Blumlein (1903–1942) was possibly the greatest British electronics engineer so far this century. By the time of his death in 1942 at the age of 38 he had been granted 128 patents – an average of one for every six weeks of his working life – testimony to an inventive genius who made a major impact on the fields of telephony, electrical measurements, sound recording, television and radar.

Mr I. Schoenberg, General Manager of EMI, said of Blumlein shortly after his death:

There was not a single subject to which he turned his mind that he did not enrich extensively.

A further series of contributions on A.D. Blumlein's life and work appears in the IEE publication *The Life and Work of A.D. Blumlein*. [5]

8.8 Digital modulation and sound radio broadcasting

In 2000 consideration began on the possibilities of digital modulation – as compared with frequency modulation – for sound radio broadcasting. The reasons for so doing were clear:

- high quality transmission, free from noise and interference, e.g. from other transmitters;
- there would be room for more radio carriers in a given band of the radio-frequency spectrum;
- the possibility of multiplexing several separate audio channels on a single carrier radio wave.

The sound radio carriers could be transmitted from terrestrial or satellite transmitters either separately or in association with television transmissions.

In the UK, USA and Japan interested broadcasters, equipment manufacturers and business men set up study groups to explore the possibilities. One such group concluded: [6]

Digital radio may not have attracted the same media attention as digital television so far, but it has the potential of a multi-million dollar business, with major implications across the range of multimedia activities.

Whilst proposals were being made to exploit digital radio commercially, the BBC announced in 2001 its plans to produce five new national radio services, including specialised music, talks, sports and news channels, and the BBC World Service. The new services were being made available in digital format via terrestrial radio, the Internet, satellites and television cables. By 2001 some 60 per cent of the population of the UK could receive these services. [7]

References

1　Asa Briggs, *The History of Broadcasting in Great Britain: Vol. 1, The Birth of Broadcasting* (Oxford University Press, 1961).
2　W.J. Baker, *A History of the Marconi Company* (Methuen and Co. Ltd., 1970).
3　P.P. Eckersley, *The Power Behind the Microphone* (Jonathan Cape, London, 1941).
4　R.W. Burns, *The life and times of A.D. Blumlein* (IEE, London, February 2000).
5　R.W. Burns, A.C. Lynch, J.A. Lodge, E.L.C. White, K.R. Thrower and R.M. Trim, 'The Life and Work of A.D. Blumlein' (IEE, *Education, Science and Engineering Journal*, London, June, 1993).
6　*Screen Digest*, 'The Prospects for Digital Radio' (2000). www.screendigest.com/rep-digrad.htm
7　BBC Digital Radio Services (2001). www.bbc.co.uk/digital radio

Chapter 9

Pioneers of television broadcasting

9.1 The birth of electronic television

When one contemplates today's superb high-quality colour television pictures –
with the prospect of even higher definition, large-screen receivers offering three-
dimensional picures to come – it is difficult to realise how crude were the early
attempts to achieve 'viewing at a distance'. It seems very doubtful whether even
the most optimistic and far-sighted of the pioneers had any clear vision of what,
in the fullness of time, was to be achieved.

The history of the development of television is a fascinating story of numerous small advances by many individuals towards the ultimate goal, some
brilliant ideas by a few creative geniuses which can now be seen to be of vital
and continuing importance, and the occasional pursuit of a blind alley of
development with all the heart-break that entailed.

The names of the earliest pioneers are now almost forgotten – one such was an
Italian priest, Abbe Caselli, who in 1862 achieved sufficient success in transmitting hand-written messages and drawings over telegraph lines to attract the
interest of Napoleon. And in 1881 an Englishman named Bidwell demonstrated
an 'electric distant vision' apparatus in which the picture to be transmitted was
analysed by a selenium light-sensitive cell moved up and down across the picture
by a cam.

In 1908 Bidwell wrote a letter to the scientific journal *Nature* proposing picture transmission by the use of 1,000 photo-electric cells, each linked to a separate light source in the receiver. Clearly a system requiring 1,000 wires was not
practical and means were required for conveying all the picture information on a
single wire. This had been realised long before Bidwell's 1908 letter and had
given rise to the concept of 'scanning', that is the examination of the picture by a
moving point in a series of close spaced lines.

9.2 Picture analysis by mechanical scanning

Paul Nipkow (1860–1940)

A little known German inventor, Paul Nipkow, is credited with the invention in 1884 of the principle of mechanical scanning based on a rotating disc with a spiral of holes. As each hole moved across the picture a line was scanned, the successive line scans falling one below the other until the whole picture had been traced. It was Nipkow's mechanical scanning system that was resurrected more than 40 years later by John Logie Baird and used for a short while by the BBC in the world's first public television broadcasting system in 1936.

9.3 Picture analysis by electronic scanning

A. Campbell-Swinton (dec. 1930)

The Scottish electrical engineer A. Campbell-Swinton patented in 1911 an electronic-scanning picture transmitting and receiving system based on cathode ray tubes, which has been described as 'an amazing piece of scientific clairvoyance, comparable perhaps to Charles Babbage's anticipation of the principle of the computer'. [1] It comprised the basic principles on which modern television cameras and receivers work.

Campbell-Swinton's early career had included activities as diverse as X-ray photography and introducing the young Marconi to Sir William Preece of the British Post Office. He was also familiar with von Braun's work in Germany on the use of the newly-invented cathode-ray tube to display the waveform of alternating currents. He had noted Bidwell's 1908 letter in *Nature* on the latter's proposal concerning the use of 1,000 photo-electric cells and 1,000 wires as a picture transmitting system and looked for means whereby the whole picture information could be transmitted over a single wire. His 1911 patent disclosed a camera tube in which a cathode-ray beam of electrons scanned a mosaic of photo-cells, line by line. The scanning beam in effect switched on each photocell in turn, releasing the charge that had been accumulated due to the light falling on that cell for the whole time since the previous scan – a pre-requisite for high sensitivity. A similar and synchronised scanning beam at the cathode-ray tube receiver generated the reproduced picture.

Campbell-Swinton's ideas for a wholly electronic television system were ahead of their time, as he himself realised; he wrote modestly:

It is an idea only, and the apparatus has never been constructed. Furthermore it could not be got to work without a great deal of experiment and probably much modification.

A major step towards the practical realisation of his proposal, which he described in his Presidential address to the Rontgen Society in 1911, was ultimately achieved by Vladimir Zworykin in his 1923 'iconoscope' patent.

Campbell-Swinton was not alone in visualising a cathode-ray tube receiver –

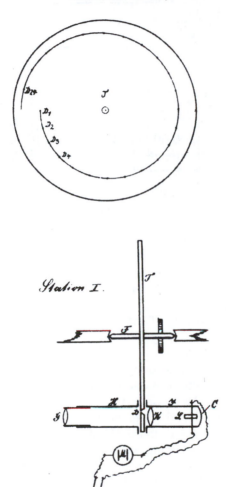

Figure 9.1   Paul Nipkow's mechanical scanning disc of 1884

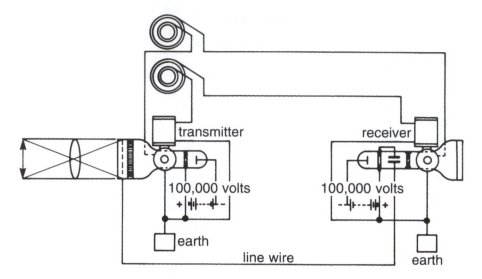

Figure 9.2   Campbell-Swinton's electronic television system of 1911

in this he had been anticipated by Professor Boris Rosing at the Technological Institute in Petrograd. However, Rosing's experimental equipment used mechanical scanning for the transmitter and, for lack of suitable amplifiers for the weak photocell signals, achieved only a faint image.

Vladimir Zworykin (1889–1982)

One of Rosing's students was a young physicist, Vladimir Zworykin, who had also studied under the famous physicist, Paul Langevin, in Paris. During the First World War he served in the Russian Army Signal Corps and later joined the Russian Wireless Telegraph and Telephone Company and began to think about the problems of 'seeing at a distance'. Perhaps inspired by his contact with Rosing, he came to the conclusion, independently of Campbell-Swinton, that electronic scanning – in which the virtually inertialess electron beam of a cathode-ray tube replaced the necessarily speed-limited rotating disc or mirror-drum of mechanical scanning – was essential for high-definition television. Furthermore it offered the prospect of continuing improvement in picture definition as the supporting technology improved.

Zworykin left Russia in 1919 after the Revolution to join the Westinghouse Electric and Manufacturing Company in Pittsburg, USA. The firm was not, however, receptive of his ideas; in spite of this he continued working on electronic scanning in his spare time and in 1923 filed a patent for his 'iconoscope' which embodied the all-important concept of electric charge storage between scans. A rival electronic scanning system developed by P. Farnsworth in the USA, which he described as an 'image dissector', did not embody the principle of charge storage with its sensitivity advantage.

Zworykin joined the Radio Corporation of America and led their television

Figure 9.3 Zworykin and the iconoscope

research. He gave a demonstration of electronic scanning to the Institute of
Radio Engineers in New York in 1931, and a paper on this subject to the Institution of Electrical Engineers in London, in 1933. In England the firm of Electric
and Musical Industries Ltd, which had links with RCA, began its own
independent development of a camera tube – the 'Emitron' – based on iconoscope principles. It was this camera tube that enabled the BBC to produce
pictures of outstanding quality right from the beginning of its public television
broadcasting service in 1936. [2]

Electronic scanning became important, not only for television broadcasting,
but also for the cathode-ray picture tubes used in visual display units (VDUs)
embodied in computers and information access (Internet/World Wide Web)
viewing systems.

Figure 9.4   John Logie Baird

9.4 A lost battle for television broadcasting

John Logie Baird (1888–1946)

John Logie Baird was a largely self-taught, highly prolific inventive genius who pursued with immense drive and confidence an approach to the design of a television broadcasting system that was nevertheless doomed to ultimate failure through its reliance on mechanical scanning principles. His efforts nevertheless had great value – above all they created an early public interest in, and awareness of, the possibilities of television as a public service, and it undoubtedly spurred on a sometimes reluctant BBC to take television seriously.

He was the son of an impoverished Scottish minister of religion; he suffered from ill health for most of his working life. His chief hobbies as a young man were experimenting with electricity and photography. He began studies at Glasgow University but these were interrupted by the outbreak of the First World War. In the post-war years he found great difficulty in getting a job in Britain and went off to the West Indies. There he set up small factories manufacturing soap and jam, but his businesses failed and he returned to England in 1922 broken in health and spirit. His biographer has written:

. . . it was an unimpressive setting for one of the greatest developments in the history of invention.

Realising he would never be a success as a trader, Baird turned his thoughts to an earlier dream derived from his hobbies of electricity and photography – the transmission of pictures over wires. He knew that photocells would generate an electric current when excited by light, and that the newly invented thermionic valves (Chapter 5) could amplify weak currents to a useful level.

Furthermore he was aware of the Nipkow disc with its spiral of holes and saw this as a possible means for scanning a picture. In 1925 Baird set up, in an attic in Frith Street, Soho, London, a 30-line scanning disc camera and a receiver with a neon tube as a light source. For a picture source he used the head of a tailor's dummy – and also, at one stage, a very scared and reluctant boy from a neighbouring office! With this improvised equipment, made from scrap material, he was able to transmit crude pictures, small in size and not much more than black silhouettes against a red background, from one side of his laboratory to the other.

Clearly much more needed to be done; better technical resources and capital were required. He approached the Marconi Company but since his apparatus had no patent protection, because of its dependence on the Nipkow disc, the company was not interested.

Partly to stimulate public interest and attract capital Baird embarked on a series of demonstrations and took out patents involving the televising of subjects in darkness by infra-red rays, colour television using a triple-spiral Nipkow disc, stereoscopic television and even the recording of picture signals on a gramophone disc.

Needless to say, with the limited technology of the time, the pictures were of crude quality.

Figure 9.5   Baird's 1926 'Televisor'

The black and red picture was barely one inch in size and had to be viewed through a magnifying lens.

However, financial backers were eventually found, attracted no doubt by Baird's demonstrated inventiveness and drive, and the Baird Television Company was formed. By 1926 a 30-line Baird 'Televisor' had been put on the market, and kits of parts were also available for home assembly.

At this stage Baird knew he had to persuade the BBC to transmit his 30-line television signals on its medium-wavelength sound broadcast transmitters. At first the BBC was not interested, understandably in view of the poor quality of the pictures. Eventually, after some persuasion by the then Postmaster General Lees-Smith, the BBC agreed that regular short public television transmissions could start from London in September, 1929. This was a great day for Baird and a series of broadcasts followed, including a transmission of the finish of the Derby in 1931. By this time he had replaced his spinning disc camera by a rotating mirror drum camera, making some improvement in picture quality, but limited by 30-line definition.

Whilst EMI and the Marconi Company were pursuing the development of a high-definition television system based on electronic scanning and the 'Emitron' camera, the Baird Television Company struggled on to improve the definition of their mirror-drum camera, first to 120 lines and then to 240 lines.

9.5 The world's first high-definition public television broadcast service is born

A marriage of the camera skills that had been developed in EMI with the VHF transmitter and aerial design know-how in the Marconi Company was formally agreed in March 1934 with the formation of the Marconi-EMI Television Company. [3]

The timing was fortunate since the trial period allowed for the Baird 30-line system had come to an end and the Government had set up a committee under Lord Selsdon to advise the Postmaster General on the future of television broadcasting in the UK.

When in 1935 the Selsdon Committee produced its report it recommended a high-definition system with not less than 240 lines per picture frame. Of the possible contenders this left only Marconi-EMI and the Baird Television Company; to resolve the problem of selection it was decided that each company should provide a service from Alexandra Palace, London on alternate weeks so that a fair evaluation of each system could be made.

The Baird Company continued to pin its faith on mechanical scanning and showed great ingenuity in achieving a 240-line standard; it used a 'flying spot scanner' for studio work and an intermediate film process for use in large studios and outside broadcasts.

The Marconi-EMI Company, armed with the electronic scanning 'Emitron' camera, elected with great courage and foresight to go to a higher definition 405-line standard. This choice was largely made possible by the brilliant research team at EMI led by Isaac Schoenberg, and which included Alan Blumlein

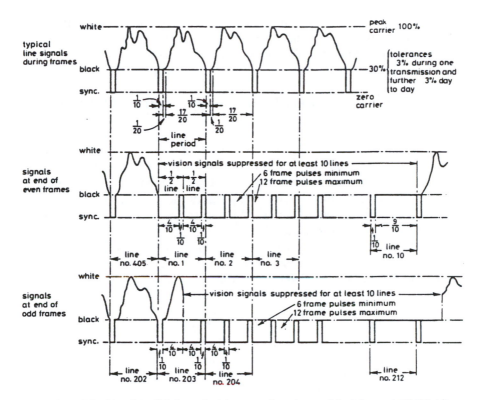

Figure 9.6   The Blumlein 405-line television waveform (as used by Marconi-EMI Ltd)

(whose contribution of stereophony to sound broadcasting is referred to in Chapter 8).

9.6 The evolution of the 405-line television standard waveform

Alan Blumlein (1903–1942)

A satisfactory television signal waveform has to include features that readily enable frame and line synchronisation, and alternate field interlacing, to be achieved in receivers; the line frequency should desirably be achieved by frequency multiplication from the UK electricity mains frequency of 50 Hz. In particular the lock to the mains frequency helped to avoid moving 'hum' bars in the received pictures.

It has been said that the manifestly successful 405-line waveform finally adopted as the British standard was created one Sunday on a breadboard at Blumlein's home, with the help of his colleagues Messrs Cork and White. The '405' came from four 3-times frequency multipliers followed by a 5-times multiplier from the 50 Hz mains frequency.

The basic 405-line waveform evolved by Blumlein and his colleagues served television development very well and stood the test of time; with appropriate modifications it later evolved into the 625-line waveform and proved capable of accommodating colour and additional services such as Ceefax/Oracle teletext.

9.7 The Baird/Marconi-EMI television system trials at Alexandra Palace

Each company provided its own vision and sound transmitters, working into a common aerial designed by C.S. Franklin of the Marconi Company. The vision transmitters operated on 45 MHz with a 4 MHz bandwidth, the sound on 41.5 MHz. The studio equipments were in striking contrast – the EMI camera being compact and mobile, whereas the Baird flying-spot scanner, and especially the intermediate film unit, were relatively immobile.

Experimental public transmissions commenced in October 1936, and the service was given a formal opening in November 1936.

The rival systems were used alternately for a time but it rapidly became apparent that the Baird flying-spot scanner, which required a studio in semi-darkness and was troublesome to the performers, was at a disadvantage compared with the Marconi-EMI 'image-orthicon' system. (The image-orthicon camera was a development from the iconoscope using electron-beam scanning and charge storage that improved both picture quality and sensitivity. [4]

Furthermore the immobility of the Baird intermediate film process and the time delay of the film processing made it unsuited to outside broadcasts.

In February 1937 the Selsdon Committee recommended that the Marconi-EMI system should carry the permanent programme service – notably on the grounds of its greater flexibility and scope for further development.

It is natural to have every sympathy with John Logie Baird who laboured unceasingly for many years, often in poverty and ill health – but it became clear that the mechanical scanning system had reached a peak of development and had no future.

An excellent and detailed history of the development of television in Britain during the formative years 1923 to 1939 has been given by Professor R.W. Burns. [5]

9.8 The television service grows and the picture improves

From the beginning in 1936, and apart from a gap in the war years, the 405-line monochrome BBC television service expanded throughout the UK, using the VHF frequency bands 41–68 MHz and 174–216 MHz. In 1953 the commercial Independent Television Service was inaugurated, and by 1964 some 97 per cent of the population could receive both programmes.

With the advent of a third service (BBC 2) in 1964 a 625-line standard was adopted and a move made to use the UHF band 470–960 MHz. The 625-line standard gave a notable improvement in picture quality, whilst the use of UHF enabled receiving aerial directivity to be sharpened compared with VHF, thereby minimising interference from unwanted transmissions and reducing radio-wave echoes from buildings, etc.

The technical characteristics of the 625-line standard were the subject of detailed study by the Research Department of the General Post Office, the GPO having a responsibility through Parliament and the Postmaster General for the regulation of broadcasting, which involved the BBC, the ITA and the UK television industry. [6] [7]

The technical standards were adopted internationally by the International Radio Committee of the International Telecomunication Union and used throughout most of Europe.

Improvements were also made in the design of camera tubes – notably the image-orthicon, and cathode-ray picture tubes for receivers, that further improved picture quality. These technical advances were perhaps not especially innovative in themselves, they were more the result of step-by-step logical development rather than highly creative new concepts. But in the latter category must come the introduction of colour to television – aptly described by G.H. Brown as 'Searching for the Rainbow'. [8]

9.9 Colour comes to television

The creation of a high-quality viable colour television system, which required a remarkable blend of sophisticated scientific study, detailed engineering design, careful economic evaluation, and eventually international agreement, may be regarded as one of the crowning achievements of 20th century technology, in which the USA, the UK, France and Germany were major participants.

As early as 1928 John Logie Baird had demonstrated a crude colour system based on a Nipkow scanning disc with three sets of spirals – in one the holes were covered by a red filter, a second by a green filter and a third by a blue filter. In 1942 Baird and an assistant, E. Anderson, experimented with a two-colour cathode-ray picture tube they called a 'telechrome', a development that ended with Baird's death in 1946 but which showed a vision beyond electro-mechanical systems.

Over the years 1940 to 1953 there was intense activity and competition in the laboratories of commercial organizations in the USA to develop a satisfactory colour television system – the prize was immense, a new mass market for tens of millions of television receivers and a massive stimulus to the advertising industry by showing its products in glowing colour.

There were two main contenders for the prize:

• the 'Columbia Broadcasting System' and its engineer Peter Goldmark with a *field-sequential* system;

- the 'Radio Corporation of America' and George H. Brown with a *dot sequential* system.

The field sequential system bore a strong resemblance to Baird's 1928 proposal in that complete red, green and blue pictures (fields) were transmitted one after the other. Whilst it had the advantage of simplicity the field-sequential system required three times the bandwidth of a black and white picture offering the same definition. If the bandwidth was reduced by lowering the field rate, problems of picture flicker arose.

RCA's proposals aimed at achieving compatability with existing black and white television receivers and fitting colour into the same video bandwidth (4 MHz in the US 525-line television standard). This created technical problems of substantial difficulty, needing great ingenuity, skill and determination for their solution.

The choice of a national colour television standard rested with the US Federal Communications Commission (FCC) which held a series of hearings of the various proposals between 1946 and 1953.

However, some of the Commissioners appear not always to have understood the subtle technicalities in the presentations made to them. And these difficulties were compounded by the intense commercial rivalry between the participants, and the feverish pace at which modifications were made to the proposed standards to overcome objections from competitors. The story of this period and the in-fighting involved has been told in illuminating detail by George Brown in his book *'and part of which I was': Recollections of a Research Engineer*. [8]

Certain highly innovative ideas, that can now be seen as crucial, emerged during this period of intense development.

9.9.1 Alda V. Bedford and 'the Mixed Highs'

The crowding of colour television into a radio channel bandwidth of 12 MHz was made possible by the principle of the 'mixed highs', developed by A.V. Bedford and adopted by RCA.

It recognised that the eye fails to distinguish colour in the fine detail of a picture, and made it possible to transmit limited bandwidth red and blue information on a 'chrominance' sub-carrier together with a full bandwidth, including green, 'luminance' channel.

In 1954 A.V. Bedford was presented with the Vladimir Zworykin Award by the Institute of Radio Engineers for his contribution to the principle of mixed highs and its application to colour television.

9.9.2 Alfred C. Schroeder: the Colour Dot Triad
9.9.3 Norman Fyler and W. E. Rowe: the Shadow Mask Tube

Earlier demonstrations by RCA to the FCC had involved the use of separate red, green and blue picture tubes and the super-position of these images by

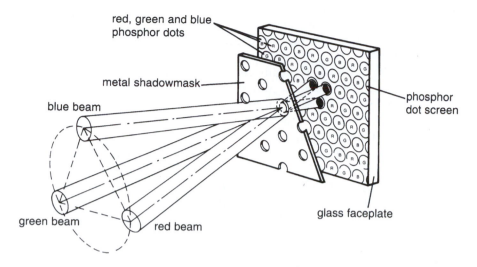

red, green and blue phosphor dots

metal shadowmask

blue beam

phosphor dot screen

green beam   red beam

glass faceplate

Figure 9.7   Principle of the RCA shadow-mask colour picture tube

dichroic mirrors – an arrangement that had registration problems and was too bulky and costly for a mass-produced domestic television receiver.

A.C. Schroeder had been working with Bedford and Kell of RCA on the problems of colour television and developed the idea of a single colour tube with groups of three red, green and blue light-emitting phosphor dots deposited directly on the inside end face of the tube. Three beams of electrons were to be aimed at the phosphors through holes in a metallic grid, arranged so that one beam struck a green dot, another a red and a third a blue dot. Schroeder filed a patent for this idea in 1947.

Fyler and Rowe developed the idea further by applying the phosphor dots directly on the inside of the end of the tube through the shadow mask, thereby automatically ensuring correct registration.

Although many variants of the shadow mask tube have since been devised, the effectiveness of the original design and its manifest suitability for manufacture in quantity undoubtedly made a major contribution to the success of colour television in the USA and Europe.

9.10 The US National Television System Committee

At the request of industry the FCC set up in 1950 a National Television System Committee (NTSC) with wide representation and powers to investigate proposals and make representations to the FCC on technical standards. In this it was assisted by the Hazeltine Laboratory which investigated the CBS, RCA and other proposals, but which took no part itself in the FCC hearings. After a detailed and massive programme of study and demonstrations to optimise the colour system performance and to secure the most economical use of the radio-frequency spectrum, the NTSC proposals were adopted by the FCC in December 1953.

According to G. Brown, the major contributions came from RCA, notably the use of a sub-carrier where hue was determined by phase and saturation by amplitude, the signal burst to synchronise colour, and the principle of mixed highs. The Hazeltine Laboratories were credited with the constant luminance principle and important contributions to the composition of the colour sub-carrier. But all in all it was an outstanding example of co-operation on the part of a large number of engineers in a joint effort to achieve the best solution to a major technical problem of common concern.

By 1979 more than 50 million NTSC-standard colour television sets, valued at $25 billion, were in use in the USA.

9.11 Colour television comes to Britain and Europe

Soon after the end of the Second World War, television engineers in the UK and Europe began following developments in colour television in the USA, and began their own investigations.

In the UK the British Post Office, with its responsibility for the regulation of broadcasting, began a detailed investigation of colour systems, including studies to define the technical parameters for optimum picture quality and efficent use of the radio spectrum. [7]

Similar studies were carried out by the BBC and the television industry, notably by Marconi-EMI. [6]

In France the Compagnie Française de Television was working on a system which sought to minimise the effects of the transmission delay distortion which might occur on cable and microwave links, by transmitting the colour information on pairs of picture lines in sequence with a delay device to combine the two, hence the name 'SECAM' (systéme électronique couleur avec memoire). The SECAM and UK systems both used frequency modulation of the colour sub-carrier instead of the suppressed-carrier amplitude modulation in the NTSC system.

In Germany the firm of Telefunken developed a system based on alternating the colour phase line-by-line, hence the name PAL (for phase alternation by line), invented by Walter Bruch.

The problem of deciding on an international standard for colour television broadcasting was one for the International Radio Consultative Committee (CCIR), a body of the International Telecommunication Union in Geneva to resolve. Beginning in 1955 the CCIR Study Group XI held a series of meetings and witnessed demonstrations of colour systems in the USA, the UK, France and Germany. A Plenary Session of the CCIR was convened in Oslo, Norway in 1966 with the aim of reaching an international agreement.

However, by that time the various national positions had become firmly entrenched and political as well as commercial rivalries determined the outcome. The USA and Japan decided to remain with the 525-line NTSC system, the UK and Western Europe (except France) opted for 625-line PAL, whilst France and the USSR chose SECAM.

The BBC commenced its colour transmissions on the PAL system in 1967, as also did the Independent Television Authority. Fortunately, programme interchange between the various systems and standards conversion later became relatively straightforward with the invention of digital techniques and the microchip. (See Chapters 11 and 12.)

George Brown, a vigorous defender of the American NTSC system, gives a forthright view of the behind-the-scenes activities that led up to the CCIR meeting in Oslo, and a critical commentary on the difficulties of arriving at decisions by committee, in his book *'and a part of which I was': Recollections of a Research Engineer.* [8]

9.12 The recording of video signals

The early recorded television programmes were based on the use of the 'flying spot telecine' technique, in which a moving photographic film is scanned by a spot of light focused on it from a cathode-ray tube. Tele-recordings were also made by filming the picture on a high-quality television monitor. Since they involved the use of photographic film with its processing requirements, these were not highly convenient methods of recording.

Magnetic tape recordings of audio signals had been made in Germany during the Second World War, but the recording of a television signal was much more difficult since it involved frequencies several hundred times higher than for audio. An early attempt at recording video signals on magnetic tape was made by the BBC Research Laboratories at Kingswood Warren, Sussex, but it required excessive tape velocities and this limited recording time.

The first commercially successful magnetic tape video recording system was developed in the USA by the Ampex Corporation and came into use in 1956. It involved the use of a moving magnetic tape, two inches wide, with a rapidly spinning drum carrying four recording heads moving across the tape. By arranging that the spinning drum laid down transverse tracks slightly inclined to the length of the tape, each short length of the tape recorded a complete scan of one picture frame.

The first uses of magnetic tape video recording were in television studios for recording and reproducing complete television programmes of an hour or more duration. However it became apparent that, provided problems of size, ease of use and cost could be solved there was an immense market for video recording equipment for use in the home.

The challenge to produce video recording equipment for the domestic market was first taken up by Philips in Holland, who demonstrated in 1972 a domestic video recorder using 0.5 inch magnetic tape in a compact casette. This had the useful capability for wipe-out and re-use of the magnetic tape, facilitating the recording of broadcast television programmes in the home.

By the mid-1970s Sony and JVC in Japan had entered the competition and produced their Betamax and VHS systems; however, these were technically incompatible. In the event, the VHS system captured the bulk of the domestic market. [9]

The availability of pre-recorded video tape casettes, providing up to two or three hours of television entertainment, proved very attractive to domestic viewers, enabling programmes to be viewed as required.

Perhaps looking back to John Logie Baird's idea of recording still pictures on gramophone records, Decca, Telefunken and Philips in Europe, Matsushita and JVC in Japan, and RCA in America began in the mid-1970s to explore the possibility of recording video signals by microscopic mechanical impressions on cylindrical or flat circular disc plastic records – the objective being low-cost manufacture in quantity by pressing the recorded video discs, analogous to the manufacture of gramophone audio disc records.

A successful video-disc technology called 'LaserVision' was developed by Philips, Holland using a digitally-modulated laser beam to impress minute pits on a helically-scanned flat circular disc in the recording mode; in the reproducing mode another laser beam was used to scan the pits and generate a video signal. Although high-quality pictures were reproduced, the discs could not be used for home recording, unlike the magnetic tape system. [9]

Since the 1980s advances in computer technology have created compact magnetically or optically recorded 'CD' discs capable of some hours of high-quality video recording. Computer digital versatile 'DVD' discs are also available with 'read and re-write' capability, enabling video programmes to be recorded and viewed.

9.13 Higher definition television

Cable television local distribution networks and direct broadcasting from satellites to viewers' homes may minimise the bandwidth restrictions imposed by the limited space available in the UHF radio spectrum for present-day television broadcasting. There was thus the prospect of improving picture definition by increasing the number of lines per picture frame above the 625-line standard, improving the quality on larger screens to be used in the home and for public viewing.

One such approach, known as HDTV (High Definition Television), pioneered by the Japanese Broadcasting Authority NHK and Sony proposed a new TV standard using 1,125 lines at 30 frames per second, requiring new designs of transmitting and receiving equipment. The picture format proposed a 9 to 16 ratio, compared with present-day 3 to 4, and 1,250 lines compared with 625, the wider aperture being more appropriate for viewing stage and field events.

(The present-day 'wide-screen' TV receivers do not, however, incorporate 'high definition' capability'.)

Another system design approach, aimed particularly at satellite transmission, was the MAC (Multiplexed Analogue Components) system devised by the Independent Broadcasting Authority in the UK and which sought an evolutionary development from the existing 625-line, 25 frames per second standard. The MAC system minimised the effect of the higher levels of noise at the upper end of the video spectrum characteristic of transmission by satellite, where the colour information is transmitted, by sending each line of colour information in a time-compressed form immediately before its monochrome component.

At the receiver the two components were time-stretched back to their normal lengths and re-united. The time compression increases the transmitted signal bandwidth; nevertheless, the overall effect would be to enable the domestic receiving aerial to be smaller than would otherwise be required. [10]

The European Community, in a bid to by-pass Japanese dominance in the consumer electronics field, sponsored in 1986 the development of an HDTV-MAC television system.

However, the difficulties of achieving commercial viability for HDTV in the consumer mass market – and the problems of compatibility involved – have so far inhibited the development of of a viable high-definition TV system in the UK and the USA.

9.14 Advances in television broadcast distribution services

Today's remarkable advances in television broadcasting – that is the distribution of television programmes simultaneously to mass, at times world-wide, viewing audiences – has been enabled by several major advances in communication system concepts and technology.

Digital technology – replacing earlier analogue technology – has not only improved the quality of transmission by rejecting noise and interference, it has enabled more effective use to be made of the limited bandwidth available in the radio-frequency spectrum and created many more frequency channels for terrestrial radio transmitters (see Chapter 13).

Coaxial cables, buried in the ground and linking viewers' homes to television programme regional distribution points, have supplemented the 'over-the-air' radio broadcast system. Initially these operated in analogue mode with a frequency-division multiplex of some half-dozen programmes. Today digital technology has enabled a hundred or more channels to be provided on coaxial cables, including channels for telephone service, access to the Internet and the World Wide Web, and 'View-on-demand' television service (see Chapters 10, 13, 19).

Optical fibre cables will gradually replace coaxial cables linking viewers' homes and have even wider potential than coaxial cables for providing new communication services (see Chapter 17).

Millimetric radio waves, using frequencies of the order of 40 MHz and small e.g. 10cm diameter aerial dishes – although not yet exploited – could offer a convenient means for providing local area coverage.

But the major alternative to terrestrial transmitters for television broadcasting – already demonstrated by its popularity – is the use of Earth-orbiting satellites in the 'geo-stationary' mode, i.e. one enabling the aerial dish at the viewers' premises to be stationary. However, satellite systems, unlike cables, do not offer unlimited capacity for service expansion since they are ultimately constrained by their use of the radio-frequency spectrum (see Chapter 15).

By 2000 the BBC, Independent TV and Sky Television had begun a programme of digital television broadcasting providing a variety of free (BBC and ITV) services and pay (Sky) services for viewers by terrestrial transmitters, satellites and cables. These services became available to some 80 per cent of the population of the UK. [11]

A further possibility for digital television distribution is the ADSL (Asymmetric Digital Line) technique developed by British Telecom which enables a video signal to be transmitted over copper pair telephone lines from exchanges to customers' homes.

References

1 Edward de Bono (ed.), *Eureka: an Illustrated History of Invention* (Thames and Hudson Ltd, Cambridge, 1974).
2 K. Geddes and G. Bussey, *Television: the First Fifty Years* (National Museum of Photography, Film, and Television, Trustees of the Science Museum, 1986), and John Logie Baird, *Sermons, Soap and Television* (Royal Television Society, London, 1988).
3 W.J. Baker, *A History of the Marconi Company* (Methuen and Co. Ltd., 1970).
4 A. Rose, P.K. Weimer and H.B. Law, 'The Image-Orthicon – A Sensitive Television Pickup Tube', *Proc. IRE*, 34, No. 7, July 1946).
5 Professor R.W. Burns, *British Television: The Formative Years (1923–1939)* (The Institution of Electrical Engineers, History of Technology Series, Vol. 7, Peter Peregrinus, London, 1986).
6 *The History of Television: from Early Days to the Present* (Institution of Electrical Engineers, Conference Publication No. 271, November 1986).
7 W.J. Bray, *Post Office Contributions to the Early History of Television*, loc. cit., p. 136.
8 G.H. Brown, *and part of which I was: Recollections of a Research Engineer* (Angus Cupar Publishers, Princeton, New Jersey, 1979).
9 David Owen and Mark Dunton, *The Complete Handbook of Video* (Penguin Books, 1982).
10 Keith Geddes and Gordon Bussey, *Television: The First Fifty Years* (National Museum of Photography, Film and Television (UK), Philips Electronics, 1986).
11 BBC Digital Television (2001). http://www.bbc.uk/reception.

Chapter 10

The engineers of the early multi-channel telephony coaxial cable systems: the first trans-Atlantic telephone cable

10.1 The growth of inter-city telephony

Beginning in the 1920s the need for ever more telephone circuits between cities in the USA and in Europe was at first met by carrier telephone systems providing some tens of telephone circuits on pole-mounted open wires or on multi-pair wire cables, using frequency-division multiplex principles (see Chapter 6).

However, as the numbers of circuits on each pair grew, problems were encountered from crosstalk between pairs and the increasing attenuation of the signals as the line frequencies were increased. As the demand grew into hundreds and more circuits on each inter-city route, a more efficient and cost-effective solution was sought.

To the engineers of the ATT Co. must be awarded the major credit for evolving the frequency-division multiplex (FDM) analogue coaxial cable system which, together with FDM microwave radio-relay systems (see Chapter 11), became the dominant modes of inter-city transmission from the 1940s to the 1980s. From the 1980s time-division multiplex (TDM) digital systems (see Chapter 13) began to supplement analogue FDM systems and eventually displace them. [1]

The most important innovative ideas and concepts which created the technological foundation for the FDM coaxial cable system are as follows.

10.2 The coaxial cable

Theoretical studies of a transmission line embodying cylindrical inner and outer conductors, i.e. a coaxial cable, were carried out by several Victorian scientists including Kelvin, Heaviside, Rayleigh and J. Thomson. In 1909 A. Russell published in the *Philosophical Magazine* a detailed analysis that enabled the cable impedance and loss to be calculated as functions of frequency.

The first practical use of coaxial cable was in submarine telegraph systems (see Chapter 3). Early applications were also found in the feeder systems of short-wave radio aerials, notably by C.S. Franklin of the Marconi Company (see Chapter 7). Franklin derived, in 1928, the optimum ratio (3.6) of outer-to-inner conductor diameter for minimum loss. The relatively low attenuation in coaxial cables, even at high frequencies, made them also suitable for television feeders operating at VHF and UHF.

From many points of view the coaxial cable was ideally suited to the requirements of a long-distance multi-channel telephony transmission system – notably the freedom from crosstalk to and from neighbouring cables, the built-in screening from other forms of electromagnetic interference such as power lines, and attenuation/frequency characteristics which offered a wide useful bandwidth capable of accommodating large numbers of telephone channels.

To H.A. Affel and L. Espenschied of ATT Co./Bell Laboratories must go the credit for creating the concept of the coaxial-cable FDM multi-channel telephony system, for which they filed a patent in May 1929 – a patent that set the stage for one of the most useful developments in telecommunications.

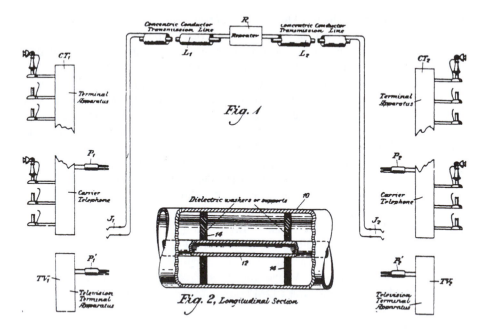

Figure 10.1   The first coaxial cable carrier telephone system patent, issued to L. Espenscheid and H.A. Affel, Bell Laboratories, May 1929

10.3 Taming the cable loss and noise

S.A. Schelkunoff and O.J. Zobel

The earlier theoretical studies of coaxial cable loss/frequency characteristics made by the Victorian scientists, mathematicians and others were extended and refined by S.A. Schelkunoff of the Bell Laboratories on the basis of electro-magnetic field theory to predict the effect of departures from an idealised geometry such as might occur in practice, and to allow for losses in insulators supporting the inner conductor. Broadly, the ohmic loss due to the surface resistance (skin effect) of the inner and outer conductors was found to increase as the square-root of the frequency, with an additional loss due to the insulators appearing in proportion to the frequency when this was sufficiently high.

The problem of equalising the variation of cable loss with frequency by suit-ably designed networks was solved by the work of O.J. Zobel and others at the Bell Laboratories. There remained the need to compensate for the overall loss of the equalised cable by valve amplifiers or repeaters – however, there were limits to the useable amplification set by valve and cable noise, and this in turn determined the maximum permissible spacing between repeaters, and thus the overall cost. [2]

J.B. Johnson and H. Nyquist

A classic study of noise in valve amplifiers due to the random nature of electron emission had been made by W. Schottky of Siemens and Halske in Germany in 1918. However, when valve noise had been reduced to a sufficiently low level, J.B. Johnson of Western Electric was able to demonstrate in 1925 that a measur-able noise contribution came from the circuit attached to the valve input. This noise was identified with thermal agitation of the electrons in the input circuit. It was H. Nyquist – born in Sweden, as was Johnson – who analysed the effect mathematically from thermo-dynamic considerations and revealed the classic formula for the noise power P in a bandwidth B:

$$P \ (noise) = 4KTB \ (watts)$$

where K is Boltzmann's constant and T is the absolute temperature (Kelvin). [3] [4]

With the problems of cable loss, equalisation and noise under control, it remained to achieve stable, linear amplification of the FDM multi-channel tel-ephony signal. Amplifier gain stability was vitally important, especially in long submarine cable FDM systems where an overall loss of perhaps 1,000 decibels (10 multiplied by itself 100 times) had to be stabilised to within a decibel (1.26 times), year in and year out. Linearity, that is a precise proportionality between the amplitudes of the output and input signal levels up to a defined overload point, is essential in order to avoid crosstalk between the telephony channels. However, the input/output characteristics of the valve amplifiers used in repeaters lacked the necessary linearity to meet the stringent requirements for high-quality, crosstalk-free transmission, especially in systems with many repeaters in tandem. (An additional requirement – phase linearity – later became evident when coaxial cable systems were used for the transmission of television signals, in order to avoid waveform distortion.)

The solution to these problems came from an outstanding invention of remarkable simplicity and power – 'negative feedback' – due to H.S. Black of the Bell Laboratories, a concept that was to have repercussions far outside telecommunications and even on economic, political and philosophical theory.

10.4 Negative feedback – a key invention of the 20th century

H.S. Black

M.J. Kelly, President of Bell Laboratories, said on the occasion of the presentation of the American Institute of Electrical Engineers Lamme Gold Medal to H.S. Black in 1957:

Although many of Harold Black's inventions have made great impact, that of the negative feedback amplifier is indeed the most outstanding. It easily ranks with Lee de

Figure 10.2　Black and a 1930 negative feedback repeater

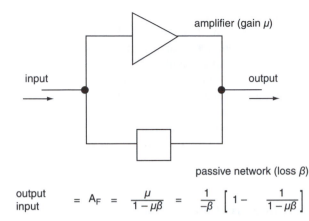

$$\frac{\text{output}}{\text{input}} \quad = \; A_F \; = \; \frac{\mu}{1-\mu\beta} \; = \; \frac{1}{-\beta}\left[1 - \frac{1}{1-\mu\beta}\right]$$

Figure 10.3   *The key invention of negative feedback, as conceived by H.S. Black of Bell Laboratories in August 1927*

Forest's invention of the audion as one of the two inventions of broadest scope and significance in electronics and communications of the past 50 years. Without the stable, distortionless amplification achieved through Black's invention, modern multi-channel trans-continental and trans-oceanic communication systems would have been impossible.

One of Black's first attempts to solve the problem in the 1920s envisaged an amplifier in which the output was reduced to the same amplitude as the input; by subtracting one from the other only the distortion products remained, these could then be amplified in a separate amplifier and used to cancel the distortion products in the original output – i.e. a 'Feed-forward Amplifier'.

Although this invention later found application in the high-power amplifiers of single-sideband short-wave radio transmitters, the need for precise balancing over a wide frequency band made it unsuitable for FDM multi-channel teleph-ony systems.

H. Black continued to wrestle with the problem for many years, without success. How it was solved is well documented: [5]

Finally, on 2 August 1927, while on the Lackawana Ferry crossing the Hudson River on the way to work, he had a flash of insight, famous in the annals of Bell Laboratories. He realized that by employing **negative** feedback, that is by inserting part of the output signal into the input in reversed phase, virtually any desired reduction in distortion could be obtained by a sacrifice in amplification. He sketched the diagram and scribbled the basic equation on a page of the *New York Times* he was carrying. He had it witnessed upon arrival at his office: negative feedback had been invented.

The key element in the invention is a passive feedback network, the loss/ frequency characteristic of which determines the overall performance of the repeater, making it substantially immune to changes of valve amplification. By suitably proportioning the gain of the amplifier and the characteristics of the network, non-linear distortion can be reduced to any desired degree.

However, the path to practical realisation was long and complex. It included a long drawn-out battle with the US Patent Office to cover the many possible applications of negative feedback – in which success was ultimately achieved with the granting of a patent in December 1937.

Much more subtle, and requiring great ingenuity, mathematical skill and clear realisation of the physical phenomena involved, was the evolution of design criteria for the passive network that enabled oscillation around the feedback loop to be avoided. Much credit for the determination of viable design criteria must go to H. Nyquist and H.W. Bode of the Bell Laboratories. [6] [7]

10.5 The power of a great idea

The applications of Black's invention and the Nyquist/Bode design criteria were not limited to repeaters for telecommunication systems – they apply to a wide range of acoustic, mechanical and electrical systems, and especially to control systems in which dynamic stability is of vital importance, e.g. automatic steering systems in aircraft and rockets. It is to the credit of the Bell Laboratories that the theory and design techniques evolved there were available world-wide.

The concept inherent in negative feedback has, at least potentially, wide practical and philosophical applications. For example, in the human nervous system the control of eye or hand movement in carrying out a task is dynamically stabilised by feedback, and can fail if the feedback is absent or distorted. The relationship between government and the governed, and between production and demand in economic systems, require adequate feedback if stability and smooth progress are to be achieved – perhaps our political and economic masters ought to look more closely at Black's invention!

10.6 Packing the telephone channels tightly: the quartz crystal band-pass filter

The economic utilisation of the useable frequency band on a transmission medium such as a wire pair or coaxial cable requires the close packing of telephone channels in the frequency spectrum – the greater the number of channels the lower is the cost per channel. For economic reasons the voice band of frequencies is customarily restricted to 200–3,300 Hz, the minimum necessary for telephonic speech of good intelligibility and satisfactory recognisability. The spacing of channels on carrier systems has been standardised at 4 kHz by international agreement via the International Telephone and Telegraph Consultative Committee of the International Telecommunication Union, as has the arrangement of blocks of channels in the frequency spectrum. The standardisation of these and other key parameters in FDM systems played a vital role in establishing telecommunication links between countries throughout the world until the

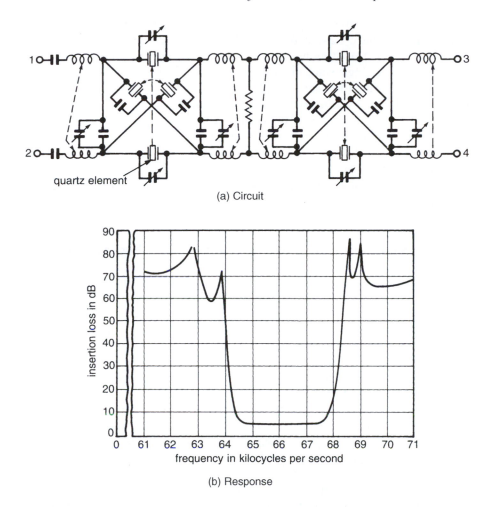

(a) Circuit

(b) Response

Figure 10.4   The channel band-pass filter using quartz crystals as circuit elements (ATT Co.)

1990s when digital time-division (TDM) systems began to be predominant (see Chapter 13).

A key factor in the design of an efficient FDM system is a bandpass filter providing low-loss transmission of the voice band 200–3,300 kHz when transposed to carrier frequencies, and the rejection of all other signals, including those in the adjacent 4 kHz wide carrier channels, by some 60 decibels (10 multiplied by itself six times) – a very stringent requirement. Whilst a solution to this problem is possible by conventional inductance/capacitance filter design using inductors with low-loss magnetic cores, a more compact and efficient design using quartz crystals as circuit elements was devised by engineers of the Bell Laboratories, the evolution of which has been described by O.E. Buckley. [8]

The very low damping (high Q factor) of the piezo-electric quartz crystal enables a very sharp cut-off to be obtained at the transmission band edges, permitting efficient utilisation of the 4 kHz channel spacing. It also facilitated the selection of a single-sideband, and rejection of the unwanted sideband, and carrier from an amplitude-modulated double-sideband signal. Carrier rejection was also facilitated by using a 'balanced' modulator.

For convenience in design of the quartz-crystal elements a basic 12-channel group, using frequencies from 60 to 108 kHz was agreed internationally via the CCITT. This formed a building block from which basic supergroups of 60 channels and hypergroups of 960 channels could be assembled, e.g. for coaxial-cable systems bearing 2,700 circuits in a 12 MHz band or 10,800 circuits in a 60 MHz band. [9]

10.7 Coaxial cable system pioneers in the British Post Office

Engineers of the British Post Office Engineering Department had maintained close contact with their counterparts in ATT Co. and the Bell Laboratories in the development of long- and short-wave radio systems from the 1920s to the 1940s, and it was natural that their attention should be drawn to the American work on the development of coaxial cable systems during the latter part of that period. In the BPO the early exploratory studies of FDM telephony and television coaxial cable systems, leading to the design and manufacture of equipment for the first field trial systems in the 1930s were carried out by a small group of engineers in the Radio Experimental Branch of the Engineering Department at the PO Research Station, Dollis Hill, in north-west London. In general, it fell to the radio engineers to carry out much of this work since their experience in the design of radio-frequency circuits and components matched the requirements of the then new coaxial-cable systems.

The engineers who carried much of the responsibility for pioneering coaxial cable system development in the BPO, under the direction of A.H. Mumford (later Engineer-in-Chief) were:

- Dr R.A. Brockbank: Negative feedback FDM repeaters;
- H. Stanesby: Quartz crystal filters;
- Capt. C.F. Booth: Quartz oscillator and filter elements;
- Dr R.F. Jarvis: Coaxial-cable and television repeaters;
- H.T. Mitchell and T. Kilvington: Television applications.

In particular Dr R.A. Brockbank and a colleague C.A. Wass made a classic study, much used in repeater design, of the loading of such amplifiers by multi-channel FDM signals and the linearity requirements to avoid inter-channel crosstalk. [10]

Their work gave a lead to the British telecommunications industry and led to a fully-engineered FDM telephony system field trial on the London–Birmingham route in 1938, using 0.45 inch inner diameter coaxial cable. [11]

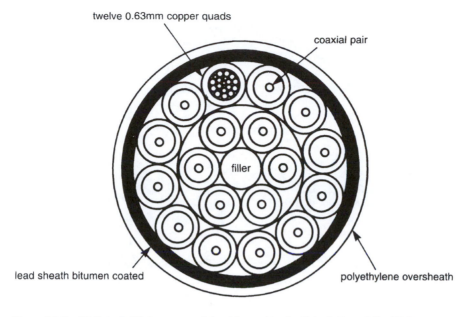

Figure 10.5   18 Pair, 2.6/9.5 mm coaxial cable used in the British Post Office/Telecom trunk network

The cable contained four coaxial cable pairs, two for FDM telephony and two for television. The telephony pairs provided a total of 400 go and return telephone channels in a frequency band between 50 and 2,100 kHz, with a channel spacing of 4 kHz. The television channel was initially limited to a bandwidth of 1.6 MHz. This was reasonably adequate for the then standard 405-line television signals but would require up-grading to meet future TV standards.

The opening in 1936 of the BBC television broadcasting service from Alexandra Palace, London created a strong public demand for television broadcasting in other cities in the UK and a corresponding pressure to develop inter-city television links. The outbreak of the Second World War in 1939 delayed further development of coaxial-cable systems until the period following the end of the war in 1945, when a growing need for more inter-city telephone circuits gave a new impetus to development.

As the years progressed from the 1940s towards the 1980s a growing family of coaxial-cable FDM telephony systems was developed, with capacities of up to 10,800 channels on each 2.6/9.5 mm coaxial pair for high-capacity routes, and also small-bore (1.2/4.4 mm) cables with up to 1,920 circuits per pair for smaller capacity routes. With this progression has come a substantial decrease – of nearly ten to one – in the relative annual cost per telephone circuit, a tribute to the far-sightedness of the pioneers who made it possible.

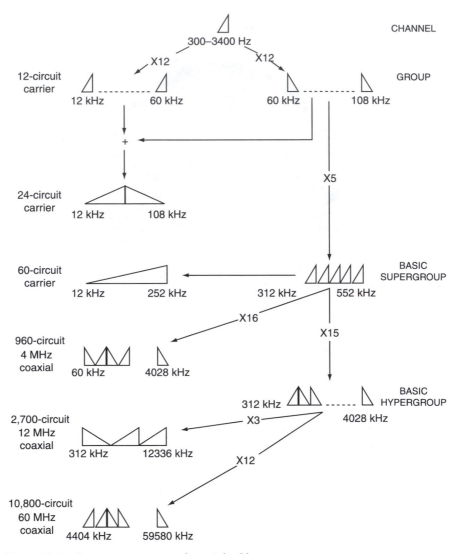

Figure 10.6   Frequency spectrum of coaxial cable systems

10.8 The first trans-Atlantic telephone cable 1956: a joint US, UK and Canada undertaking

AND WHEREAS it is desired to provide a submarine cable system between - he USA and Canada on the west, and the UK in the east . . .

These words in a formal legal agreement between the British Post Office, the ATT Co. and the Canadian Overseas Telecommunication Corporation marked the beginning of the most outstanding engineering achievement in telecom-

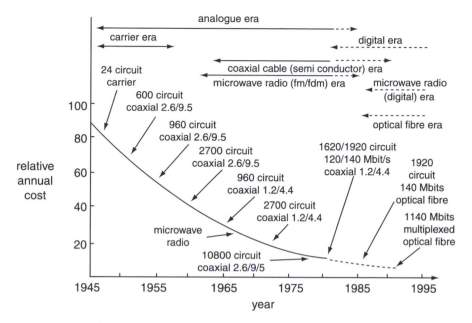

Figure 10.7   Relative annual costs (100 km)

munications during the first half of the 20th century, nearly a hundred years after the first trans-Atlantic telegraph cable. The legal agreement, and the associated contract, represented an act of faith on the part of the designers and manufacturers, on both sides of the Atlantic, to create in the short space of three years a long multichannel telephony submarine cable system that demanded higher standards of precision in the design of the cable and repeaters, and longer life-expectancy than had ever before been achieved. A detailed account by British and American engineers of the design, manufacture and laying of the system is given in the *PO Electrical Engineers Journal*, January 1957. [12]

It was a stupendous pioneering undertaking. Some 4,500 miles of coaxial cable had to be made to the most exacting specification ever devised, and new machinery had to be designed for laying the cable in waters up to 2.5 miles deep. Surveys of the Atlantic Ocean had to be carried out to select the most suitable route and 146 repeaters had to be built, capable of withstanding the rigours of laying and the extreme water pressures in the deep ocean, and able to function without attention for at least 20 years.

The main Atlantic crossing was carried out by two cables, one for each direction of transmission, with one-way repeaters in flexible housings, designed and built by ATT Co. and Bell Laboratories. This deep-sea section extended from Oban, Scotland to Clarenville, Newfoundland, a distance of about 2,000 miles.

From Clarenville a second cable system continued to Sydney Mines, Nova Scotia for 340 miles, partly in coastal waters and partly overland, using a single cable with two-way repeaters in rigid housings, of British Post Office Engineering Department design. From Sydney Mines multi-channel carrier and

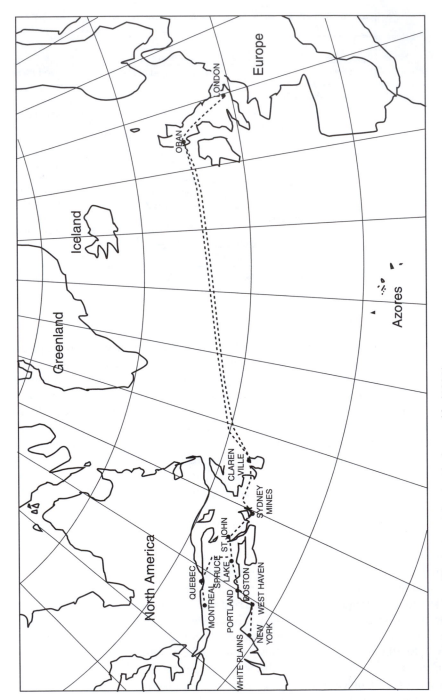

Figure 10.8   Route of first trans-Atlantic telephone cable (1956)

Figure 10.9   H.M.S. Monarch. Laid first TAT cable

microwave radio-relay systems extended the telephone circuits to New York and Montreal.

The completed system provided 35 high-quality telephone circuits from London, 29 to New York and 6 to Montreal. Clearly, a project of this complexity and magnitude involved the contributions of many scientists and engineers in the ATT Co. and Bell Laboratories, the British Post Office and UK manufacturing industry. Much credit for technical oversight and anagement of the project must go to the following:

- Dr Mervin J. Kelly, President Bell Laboratories;
- Sir Gordon Radley, Director General BPO (former Director of Research and Engineer-in-Chief);
- R.J. Halsey, Assistant Engineer-in-Chief (former Director of Research).

The technical problems to be solved were substantial. On the 2,000-mile main trans-Atlantic section the cable attenuation at the highest working frequency of 164 kHz was 3,200 dB (equivalent to 10 multiplied by itself an incredible 320 times!), which was compensated by 51 deep-water repeaters, each accommodating 36 telephone circuits. On the Newfoundland–Nova Scotia 340-mile section the cable attenuation at the highest working frequency of 552 kHz was 1,000 decibels, compensated by 16 repeaters, each accommodating 60 two-way telephone circuits. These large gains had to be stabilised, allowing for valve ageing and cable loss changes due to temperature variation and other causes, to within

a decibel or two, year in and year out. Without H.S. Black's invention of negative feedback and the further contributions by Bode and Nyquist this would
have been impossible.

The problem of achieving long life deserves special mention. Clearly, all components used in repeaters had to be carefully scrutinised and tested to ensure
lives of 20 years or more but the thermionic valves, the main active device,
presented unique problems.

Valve life, and the development of quality assurance techniques, had long
been the subject of intensive study in Bell Telephone Laboratories and in the
Research Branch of the British Post Office at Dollis Hill, London, under the
direction of Dr G. Metson. The latter work, which has been described by M.F.
Holmes, had its origins in BPO development of North Sea submarine telephone cable systems and later found important applications in trans-Atlantic
and trans-Pacific submarine cable systems.

The problems of laying the cable – and recovering it from ocean depths of a
mile or more in the event of a cable or repeater fault – were successfully solved
by the BPO Research Department design of a linear cable engine mounted in the
bow of the BPO Telegraph Cable Ship *HMS Monarch*. [12] [13]

10.9 Impact of the first trans-Atlantic telephone cable

Although the number of telephone circuits provided, 35 for commercial use, was
small compared with thousands on the later cable and satellite systems, the first
trans-Atlantic telephone cable gave a major impetus to traffic growth compared
with the short-wave radio-telephone systems, mainly through the greater reliability and better speech quality available on the cable, paving the way for higher
capacity systems in the years that followed.

The first trans-Atlantic telephone cable (TAT1) was followed three years later
by a telephone cable (TAT2) between Newfoundland and France. In 1961
another milestone was reached in the Atlantic with the laying of a 60-circuit
telephone cable from the UK to Canada (CANTAT 1). This system employed
light-weight cable in the deep water sections and both-direction rigid repeaters,
the basic design of which later became a world standard.

The light-weight cable design was originated by Dr R.A. Brockbank of the
BPO Research Department; it deserves special mention for its revolutionary
impact on submarine cable system design. Earlier cables relied for their strength
on heavy external iron-wire armouring; in the Brockbank design the strength
element was inside the inner coaxial conductor. This design was cheaper to
manufacture, easier to lay because of its lighter weight and greater flexibility,
and enabled much longer lengths of cable to be accommodated in the holds of
cable laying ships.

In the 1960s transistorised repeaters became available, enabling further
advances to be made in the performance and capacity of submarine cable
systems (see Chapter 12).

By 1976 4,000 circuit capacity had been achieved on TAT6 and by 1988 40,000 circuits on an optical-fibre trans-Atlantic cable (TAT8) (see Chapter 17).

References

1 Gene O'Neill (ed.), *A History of Engineering and Science in the Bell System: Transmission Technology* (1925–1975) (ATT Co./Bell Laboratories 1985).
2 O.J. Zobell, 'Extensions to the Theory and Design of Electric Wave Filters' (*Bell Syst. Tech. Jl.*, 10, April 1931).
3 J.B. Johnson, 'Thermal Agitation of Electricity in Conductors' (*Phys. Rev.* 32, July 1928).
4 H. Nyquist, 'Thermal Agitation of Electric Charge in Conductors' (*Phys. Rev.* 32, July 1928).
5 Robert V. Bruce, *Alexander Graham Bell and the Conquest of Solitude* (Victor Gollancz, 1973).
6 H. Nyquist, 'Regeneration Theory' (*Bell Syst. Tech. Jl.*, January 1932).
7 H.W. Bode, 'Relations Between Attentuation and Phase in Feedback Amplifier Design' (*Bell Syst. Tech. Jl.*, January 1932).
8 O.E. Buckley, 'Evolution of the Crystal Wave-Filter' (*Jl. Appl. Physics* 8, January 1937).
9 *Post Office Electrical Engineers Journal, 1906–1981*, 75th Anniversary, 74, part.3, (October 1981).
10 R.A. Brockbank and C.A. Wass, 'Non-Linear Distortion in Transmission Systems' (*Jl. IEE* 92, 1945).
11 A.H. Mumford, 'The London – Birmingham Coaxial Cable System' (*Post Office Electrical Engineers Journal*, 30 Pt. 3, October 1937).
12 M.F. Holmes, 'Active Element in Submerged Repeaters: the First Quarter Century' (*Proc. IEE*, 123 No.10R, October 1976).
13 *The Post Office Electrical Engineers Journal: The Trans-Atlantic Telephone Cable*, Number 49 Pt. 4, January 1957.

Chapter 11

The first microwave radio-relay engineers

11.1 What are microwaves?

'Microwaves' are electromagnetic radio waves with frequencies from about 1,000 MHz (30 cm wavelength) to 30,000 MHz or 30 Ghz (1 cm wavelength); sometimes called '*centimetric*' waves. Radio waves of even shorter wavelengths, from 30 to 3,000 GHz (10 mm to 0.1mm wavelength) are referred to as 'millimetric' waves. Beyond the millimetric waves are the infra-red and visible-light regions of the electromagnetic wave spectrum.

Microwaves are remarkably suitable for the purposes of point-to-point telecommunication because they can be sharply beamed by directional aerials of mechanically convenient dimensions, a property which also made them very effective for the radar systems developed during the Second World War. Since microwaves do not readily diffract around the curved surface of the Earth, they are mainly useful over 'line-of-sight' paths between transmitter and receiver. An exception to this generalisation is possible by the mechanism of 'tropospheric scattering' in which microwaves can be propagated to beyond the visible horizon by wave scattering from regions of the lower atmosphere or troposphere characterised by non-uniform humidity and temperature.

The discovery of the existence and useful properties of microwaves owes much to Heinrich Hertz and his experiments in 1886 which confirmed Clerk Maxwell's predictions of the identity of radio waves and light. Hertz's experiments with spark generators and spark gaps, resonant loops and parabolic reflectors were probably conducted on frequencies up to some 600 MHz (50 cm wavelength), nevertheless they laid the foundation for developments that were to prove of great value. Hertz did not himself envisage the use of microwaves for long-distance communication because 'it would require a mirror as large as a continent!', Chapter 2.

Hertz's work has been recognised by the usage internationally of the term 'Hertzian Waves', and by the French of the term 'Faisçeau Hertzien' to describe a beamed microwave link.

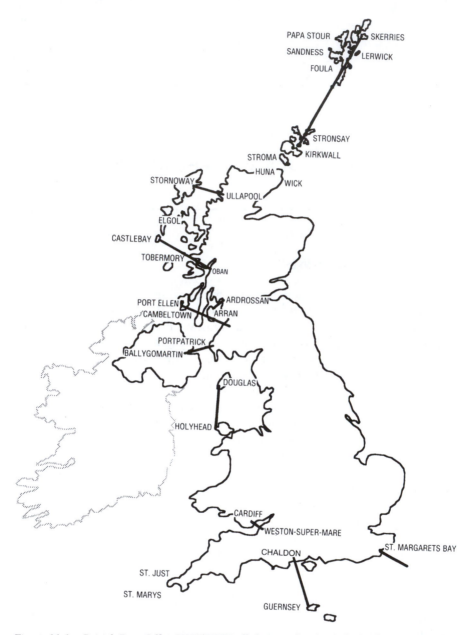

Figure 11.1   British Post Office VHF/UHF off-shore multi-channel telephony links (1938)

By the 1920s other experimenters using similar techniques had generated frequencies up to 15 GHz and beyond, but practical applications of microwaves had to await other events.

11.2 Pre-war VHF/UHF and microwave radio-relay developments

In the 1930s the British Post Office made extensive use of VHF radio links to provide single telephone circuits, or groups of 6, 12 or 24 frequency-division multiplex circuits between the mainland and off-shore islands, e.g. off the coasts of Scotland and Wales, and to the Channel Islands. These links were engineered by the Radio Experimental Branch of the Post Office at Dollis Hill, London under the direction of A.H. Mumford, H.T. Mitchell and D.A. Thorn.

Frequencies in the range 50 to 100 MHz were generally used, the lack of commercially available valves precluding the use of higher frequencies at the time. Initially the radio transmitters and receivers used 'amplitude modulation' of the carrier wave; a later development used 'frequency modulation'.

The aerials were commonly arrays of half-wavelength dipoles, or a scaled-down version of the convenient and economical Bruce rhombic aerial used in short-wave overseas services (see Chapter 7).

A 65 MHz link providing 15 telephone circuits over a 40- mile path between Scotland and Northern Ireland was brought into service by the British Post Office in 1936. The link to the Channel Islands involved an 85-mile oversea path between Chaldon, Dorset and Guernsey, Channel Islands. Frequencies of 37.5 and 60 MHz were used for the commercial telephony service, enabling valuable radio-wave propagation data to be obtained by comparing transmission over two widely different radio frequencies. [1]

This four-channel telephony link was closed down during the war years; its reinstatement after the end of the war in 1945 played a major role in restoring the morale of the Channel Islanders.

Similar developments using the VHF/UHF spectrum took place in the USA, notably by the Bell Laboratories. An early example was a single-channel two-way telephony link established between Green Harbour and Provincetown, Massachusetts in 1934, using frequencies of 63 and 65 MHz. An amplitude-modulated FDM link, providing five telephone channels, using frequencies of 156 and 161 MHz was set up on a 26-mile path between Cape Charles, Maryland and Norfolk, Virginia in 1941. [2]

These early essays in FDM multi-channel telephony operation of VHF/UHF radio links provided valuable experience and a degree of useful commercial service. However, they revealed the limitations of amplitude modulation arising from inadequate linearity of the modulation and amplification processes, which gave rise to inter-channel crosstalk. It also became clear that the frequency space available in the VHF/UHF spectrum was insufficient to allow for traffic growth and any large-scale exploitation. These considerations prompted research to find more effective modulation methods and means for using the spectrum above 1,000 MHz – microwaves.

11.3 Frequency modulation provides a solution

Major E. Armstrong's invention of frequency modulation FM had proved of great value in VHF broadcasting, notably by improving the signal-to-noise ratio as compared with amplitude modulation, and improving transmission quality by minimising non-linear distortion.

The first application of FM to frequency-division multiplex radio telephony systems arose in the UK in the late 1930s as a means of reducing noise on a 12-channel amplitude modulation VHF link on a 65-mile path between Holyhead, Anglesey and Douglas, Isle of Man. This link suffered at times from interference due to salt-encrusted insulators on a high-voltage power line near the receiving aerials. Post Office Radio Experimental Branch engineers J.H.H. Merriman and R.W. White at the BPO Castleton Laboratories, South Wales designed and built frequency modulation equipment suitable for a 12-channel FDM 60–108 kHz group. [3] The experiment proved a dramatic success; not only was the power-line interference greatly reduced, a projected increase of radio transmitter power to overcome the interference was avoided, but inter-channel crosstalk was virtually eliminated. (The writer made a small contribution to the latter by a theoretical study enabling the phase linearity requirements for low-distortion transmission of a frequency-modulated signal to be determined; a similar but more precise study was later made by L.L. Lewin of the Standard Telecommunication Laboratories, UK.) [4]

Figure 11.2   The first experimental microwave link (STC/LMT) across the English Channel (1931)

This successful use of frequency modulation for FDM telephony radio-relay systems had an impact far beyond the specific application in which it was first tried – it became the norm for microwave radio-relay systems bearing a thousand or more telephone channels and for television relaying, standardized by the International Radio Consultative Committee and used throughout the world. [5] [6]

11.4 A pioneering microwave radio-relay experiment (1931)

To the scientists and engineers of Standard Telephones and Cables Ltd, UK and Les Laboratoires le Matériel Téléphonique, Paris must go the credit for a first pioneering experiment in the use of microwaves that heralded the shape of things to come. This experiment in March 1931 involved a 1,720 MHz link over a 21-mile path across the English Channel between Calais and St Margaret's Bay, near Dover, providing a single two-way telephone channel. [7]

This link used front-fed parabolic-reflector aerials some three metres in diameter, with a power gain of 26 decibels – this large gain enabled the link to function with a transmitter power of less than 0.5 watt. The transmitter consisted of an oscillator invented by Barkhausen-Kurz in 1919 – a triode valve in which the grid was held at a high positive potential and the anode at a low one, the electrons oscillating between grid and anode at an extremely high frequency, largely independent of the external circuit. It is not clear whether the modulation was in amplitude or frequency, perhaps a combination of both.

The experiment was notable not only for its pioneering use of microwaves, but also because it revealed propagation variations caused by the rise and fall of the tide in the English Channel and its effect on the phase of the sea-reflected wave – a phenomenon that had to be allowed for in the design of the first commercial cross-channel microwave link in the 1950s.

11.5 The impact of the war years on microwave development

The war-time development of radar – the fascinating story of which is outside the scope of this book – gave an impetus to the evolution of microwave techniques and devices in many areas including aerials and waveguides, klystron oscillators, pulsed magnetrons, silicon crystal mixers, microwave filters and other components, and intermediate frequency (typically 45 MHz) valve amplifiers.

The klystron was one of the most valuable contributions to microwave radio-relay system development. It was invented by the brothers R.H. and S.F. Varian of Stanford University, USA in 1939, using the principle of velocity modulation of an electron stream by a varying applied voltage generated across a gap in a microwave cavity resonator. Velocity variation caused electrons to 'bunch', faster electrons overtaking slower ones, as they progressed along the stream.

Energy could be abstracted from the bunches by a second suitably placed

microwave cavity resonator, thus creating a microwave amplifier. Similar principles could be applied in a single cavity 'reflex' system, creating a microwave
oscillator. [8]

The war also stimulated the development of time-division multiplex telephony techniques using width or position modulation of trains of pulsed
microwave carriers generated by klystrons or magnetrons, foreshadowed in
their 1938 patent by E.M. Deloraine of LMT in France and A.H.Reeves of
STC in England [9]. In England a pulse-modulation microwave equipment
providing 8 telephone channels for military use was brought into service in
1942; it used a 5 GHz magnetron. In Germany a 10-channel system
designed by Lorenz used initially a 1.3 GHz magnetron and later a 1.3 GHz
klystron. [5]

As the war-time development of radar moved up the spectrum from 3 GHz
(10 cm–'S' band) to 10 GHz (3 cm–'X' band) the need for low-loss transmission
of microwave power stimulated the evolution of waveguides – generally in the
form of rectangular copper tubes with dimensions related to the wavelength.
Coaxial cables had too high a loss at such frequencies, and open-wire lines
radiated too freely. At microwave frequencies conventional circuit elements such
as inductors and capacitors became vanishingly small, and here waveguide
components became available to take their place in filters and power dividers/
combiners.

Lord Rayleigh had shown in 1897 that certain solutions of Maxwell's equations (see Chapter 2) predicted that the transmission of electromagnetic waves
through hollow conducting tubes was feasible, and that a variety of modes
involving various combinations of magnetic and electric field patterns was
possible. Pioneering work in Bell Laboratories in the 1930s, notably by G.C.
Southworth, J.R. Carson, S.P. Mead and S.A. Schelkunoff, established the basic
criteria for waveguide design, including the cut-off frequencies and losses due to
imperfect conductivity. At Massachusetts Institute of Technology W.L. Barrow
was also conducting a programme of research on waveguides, including the use
of resonant waveguide cavities as filters.

Thus, by the end of the war in 1945 the ground-work had been prepared for
the development of the high-capacity microwave radio-relay systems that were
to become of major importance in the civil telecommunication networks of the
USA, Europe, Japan and other countries.

11.6 Post-war microwave radio-relay system research in the Bell Laboratories, USA

System studies by the ATT Co. in the later years of the war had indicated
that there was a good possibility that microwave radio-relay systems could
provide an economic alternative to coaxial cable systems for both multi-
channel telephony and television, with the advantage of speedier provision in
some cases. A substantial programme of research and system development

was initiated in Bell Laboratories on the basis of this forward-looking study. [2]

Central to the research programme were studies of microwave propagation over typical land and sea line-of-sight paths, with the objectives of determining the optimum region of the spectrum for radio-relay system use and the rules for system design, recognising that microwaves were subject to a degree of fading under certain weather conditions and that attenuation due to heavy rainfall could affect the higher microwave frequencies. These studies indicated the suitability of frequencies around 4 GHz (7.5 cm wavelength) for the first systems to be built, with the possibility of later using other frequencies in the range 2 to 10 GHz. An important consideration in system planning was the necessary clearance of the direct ray path over intervening terrain to avoid excessive fading due to the ground or sea reflected rays and ray-bending from abnormal atmospheric refraction – economically important because it governed the height and cost of towers needed to support the microwave aerials. The propagation studies of microwave fading had to be carried out over periods of a year or more to provide statistical data enabling microwave repeater design to incorporate sufficient reserve gain and signal-to-noise margin to maintain the necessary overall standards of performance of long radio-relay systems.

Concurrently with the work on microwave propagation, the Bell Laboratories at Holmdel, New Jersey were pursuing fundamental studies of microwave components, aerials, repeater and system design under the leadership of Harald Friis.

Harald T. Friis (1893–1968)

It is said of H.T. Friis that:

... Friis, an outstanding researcher and radio engineer, in addition to innumerable personal contributions, served as mentor to a generation of Bell Laboratories radio scientists and engineers.

He was as much loved and respected for his human qualities as for his technical acumen. To those who knew him, the enormous microwave radio-relay network is a fitting memorial to his work. [2]

He has told his own life-story in a fascinating book *Seventy-five Years in an Exciting World* [10]. He was born in a small town in Denmark in 1893, one of a family of ten children, four of whom died in infancy. His father was a brewer who died when 48 years old, leaving his mother with six children and a heavily mortgaged brewery to run. Family life was conducted in very straightened circumstances, and Harald was apprenticed to a blacksmith for a time. During his technical college studies he was fortunate to study under Professor Niels Neilson, the famous mathematician; he also worked under Valdemar Poulsen, the inventor of the Poulsen Arc and a magnetic-tape system for recording speech.

Friis joined the Bell Laboratories in 1919 – one of his first tasks was to work on a 'double-detection' radio receiver, i.e. one in which an incoming modulated radio-frequency carrier is first translated to a, generally lower, intermediate frequency for convenient amplification and before demodulation – a principle which he describes as 'one of the most important contributions to the radio art', attributed to Lucien Levy in France. His early work was on long-wave reception, moving on later to short-wave radio, including single-sideband and multiple-unit steerable antenna (MUSA) systems (see Chapter 7). He had much to do with the setting up of the Holmdel field laboratory and the creation of a down-to-Earth, informal environment in wooden huts. The writer visited Holmdel shortly after the end of the war, at a time when the BPO Engineering Deptart-ment was beginning to plan new laboratories, and I asked Friis 'how big did he think an ideal research laboratory should be?'. Harald thought for a while and said 'Perhaps about 100 people, I could then hand-pick them for the jobs to be done'. Today's laboratories with staff measured in thousands, housed in air-conditioned, glass-fronted multi-story buildings, are a long way from Friis's ideal!

Not even Holmdel was immune to visits from high authority; on one occasion the writer happened to be in the small conference room in one of the wooden huts and noticed that, on a blackboard full of scrawled circuit diagrams, math-ematical symbols and triple-integrals, a firm clear hand had written 'dollars per message channel per mile' – on enquiry I was told that Dr Kelly, then President of the Laboratories, had just paid a visit, and evidently determined to make clear to his scientists/engineers what their real objective was!

In the field of microwave radio Friis will be remembered in particular for two important contributions. One stated a law, as important for microwave system design as Ohm's Law is for electric circuits:

$$P_r/P_t = A_t A_r / \lambda d$$

where P_r and P_t are the received and transmitted powers, A_t and A_r are the

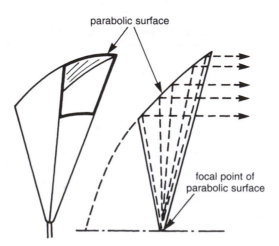

parabolic surface

focal point of parabolic surface

Figure 11.3   The Friis–Beck horn-reflector aerial

effective areas of the transmitting and receiving aerials, λ is the wavelength, and d is the distance between the transmitting and receiving aerials.

All good microwave engineers should have this equation firmly in their heads if not engraved on their hearts!

The second notable contribution was the 'horn-reflector aerial' invented by Friis and A.C. Beck at Holmdel, and based on earlier work by G.C. Southworth on the radiating properties of conical horns. [11]

The horn-reflector aerial has a number of valuable properties; it is effective over a wide frequency range, typically from 2 to 10 GHz, in a single aerial, with good impedance matching over the whole of this range, enabling it to be used simultaneously for a number of groups of channels on widely different frequencies. Also it can accommodate both vertical and horizontal polarisation, and it offers good protection against unwanted signals from the side and behind, an important feature in multiple-hop radio-relay system design.

So successful was the horn reflector aerial that it was used widely for radio-relay systems throughout the USA, in Europe and in other parts of the World. A steerable version at Holmdel was used for wave propagation research and became a prototype for the much larger radome-protectedm steerable horn-reflector aerial for the Telstar communication satellite experiment at Andover, Maine in 1961 (see Chapter 15).

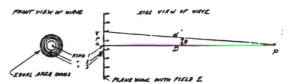

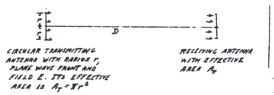

Figure 11.4 Derivation by H.T. Friis of the fundamental microwave transmission formula

$$Pr/Pt = \frac{At \times Ar}{\lambda^2 \times d^2}$$

Of Harald Friis an eminent colleague J.R. Pierce has written in the preface to Harald's Memoirs:

... I have known cleverer inventors, more abstruse scientists, deeper mathematicians, better politicians, and executives of higher degree. I have known no other man who has left as deep and profitable impress on those who have worked for or with him, or who has had a clearer insight or a surer success in the work he has undertaken. [10]

11.7 Microwave radio-relay system development by ATT Co./Bell

From about 1945 microwave radio-relay system development in the USA received a strong impetus from the urgent need of the ATT Co.mpany to provide television relaying facilities throughout the USA to meet the needs of the powerful broadcasting companies – the Columbia Broadcasting System, the National Broadcasting System and the Radio Corporation of America. The economics of television programme provision, and the needs of the advertisers on whom the whole financial structure depended, required the ability of each broadcasting company to link its many television stations so that a single programme or advertisement could be transmitted throughout the nation. Any failure by ATT to provide high-quality, reliable nation-wide television relaying facilities in good time would have seriously jeopardised its virtual monopoly over long-distance inter-state communication by opening the door to competitors.

The ATT Co. already had a nation-wide network of coaxial cables, each transmitting a thousand or more telephone circuits. However, the coaxial cable network had been designed for maximum efficiency in the transmission of multi-channel FDM telephony and was not technically well-suited to television relaying without considerable modification.

Fortunately – or by good foresight – the microwave research at Holmdell had reached a point where a multi-hop radio-relay field trial could be planned, built and tested. The route chosen was from New York to Boston, a distance of 220 miles, in five hops averaging 27.5 miles. The frequencies used for two microwave carriers in each direction of transmission were in the 3,700–4,200 MHz band eventually assigned by the US Federal Communications Commission for 'common-carrier' use. Frequency modulation of the microwave carriers was used, both for television and for FDM multi-channel telephony, the primary emphasis of the experiment being on television. The aerials on this experimental system – designated 'TDX' – were of the 'delay-lens' type designed by Dr Koch of the Bell Laboratories. 'Waveguide branching' units, invented by W. Tyrell of the Laboratories, enabled a number of microwave carriers on different frequencies to be combined for transmission via a common aerial, or separated for reception.

Repeater design was based on frequency conversion of the incoming microwave carrier, via a crystal mixer and klystron local oscillator, to an intermediate frequency of 65 MHz at which amplification and gain regulation to allow for fading took place.

The intermediate frequency signal was then converted in a crystal mixer to an outgoing microwave carrier displaced 40 MHz from the incoming carrier frequency. This frequency shift was necessary to avoid undesirable feedback from the output to the input of the repeater via the limited back-end directivity of the transmitting and receiving aerials. The strength of the outgoing microwave carrier was increased in a velocity-modulation klystron amplifier to a power of about 1 watt.

Overall performance tests in 1947 demonstrated that television picture quality was virtually unimpaired after transmission over the TDX system, and some 240 FDM telephone circuits of satisfactory quality were possible [11].

The basic design concepts of the TDX experimental microwave radio-relay system were demonstrably sound and were followed in the TD2 operational system that was eventually used by ATT Co. throughout the USA for televison and multi-channel FDM telephony.

Initial changes in design incorporated in the TD2 system included the use of the Friis horn-reflector aerial in place of the delay-lens aerial, and a 4 GHz triode valve power amplifier in place of the klystron amplifier, with the advantages of lower operating voltages and longer life. [13]

The 4 GHz triode, designed by Dr J.B. Morton of the Bell Laboratories, took the upper frequency limit of operation of the triode valve to a point much higher than had earlier been achieved. It was a remarkable tour-de-force, requiring extremely fine mechanical tolerances on the spacings between the cathode, grid and anode, and precise control of the manufacturing processes.

Over the period 1950 to 1980 a series of design improvements were made in the TD2 4 GHz radio-relay system, and eventually 1,800 FDM telephone circuits were achieved on each radio carrier, and up to 11 carriers on a route. Microwave systems on frequencies of 6 GHz (TH) and 11 GHz (TL) were later developed.

These systems all used frequency modulation, but some later developments explored the use of single-sideband amplitude modulation for FDM telephony, the smaller radio-frequency bandwidth compared with frequency modulation offering significant gains in efficiency of radio spectrum utilisation.

The demonstrated success of the TDX experimental microwave radio-relay system and its operational successors stimulated similar developments in the UK, Europe and Japan, with variations of design according to the various national requirements and resources.

11.8 Microwave radio-relay system development in the UK

When the study of radio-relay by the Radio Experimental Branch of the British Post Office Engineering Department was resumed shortly after the end of the war there was a lack of commercially available microwave amplifiers in the UK, and attention was first directed to the use of frequencies around 200 MHz for which triode valve amplifiers with power outputs of a few watts could be obtained.

An experimental radio-relay system operating at 200 MHz was designed and built by the Radio Branch of the ED between the BPO Research Station at Dollis Hill, London and ED Laboratories at Castleton, South Wales. This served primarily as a test-bed for radio-relay system experimental work but was spurred on by the need for a temporary television link between London and Wales for the opening of the British Broadcasting Company television transmitter at Wenvoe, near Cardiff in 1952.

This five-hop radio-relay system was unique in that it was designed to use a single frequency throughout, with Bruce-type horizontal rhombic aerials in pairs on opposite sides of hill-top sites, so located that diffraction loss over the hill-top avoided interaction between the outgoing and incoming 200 MHz signals.

The London–Wenvoe system was also of interest in that it marked one of the earliest uses of frequency modulation for the radio-relaying of television signals in the UK. It demonstrated clearly the advantages of frequency modulation over amplitude modulation for such applications, and this became the accepted type of modulation for future radio-relay systems. [14]

A 900 MHz radio-relay system for television was built under Post Office contract by the General Electric Company (UK) to link the television switching centre at the PO Museum Exchange in London with the BBC television transmitter at Sutton Coldfield in Birmingham, and opened in 1949. [15]

By the 1950s, and with an eye on developments in the USA, it had become clear that the future development of radio-relaying, both for multi-channel telephony and for television, lay in the microwave region of the radio spectrum where sufficient frequency space was available for large-scale exploitation. The first commercial development of microwave radio-relay systems in the UK was carried out by Standard Telephones and Cables Ltd with their 4 GHz system, using a travelling-wave tube as a transmitter output stage power amplifier, and by the General Electric Co (UK), with a 2 GHz system using a disc-seal triode valve as an output stage.

The STC 4 GHz radio-relay system, built under contract to the British Post Office for the Manchester–Kirk o'Shotts, Scotland television link and opened for service in 1952, was the first microwave radio-relay system in the world to use travelling-wave tubes (TWT) on an operational basis. [16]

The BPO Engineering Department, recognising the design and manufacturing problems of triode valves with close-spaced electrodes, had earlier arranged sponsorship of TWT development with STC via the UK Co-ordination of Valve Development Committee. This development work, directed by D.C. Rogers of STC at their Ilminster, Somerset laboratories, resulted by 1949 in the highly successful one-watt output 4 GHz travelling-wave tube amplifier first used in the Manchester–Kirk o'Shotts radio-relay system.

The travelling-wave tube proved of major importance in the development of microwave radio-relay systems – it offered wide signal bandwidth and could be

(a) Rudolph Kompfner (RH) and John Pierce (LH)

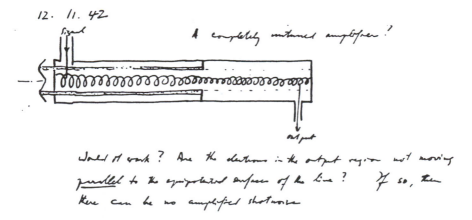

Notebook entry dated Nov. 12, 1942. Output resonator is replaced by section of helix, leading to astonishing conclusion: "A completely untuned amplifier?"

(b) From Rudolf Kompfner's notebook

Figure 11.5   Pierce, Kompfner and travelling wave tube notebook entry

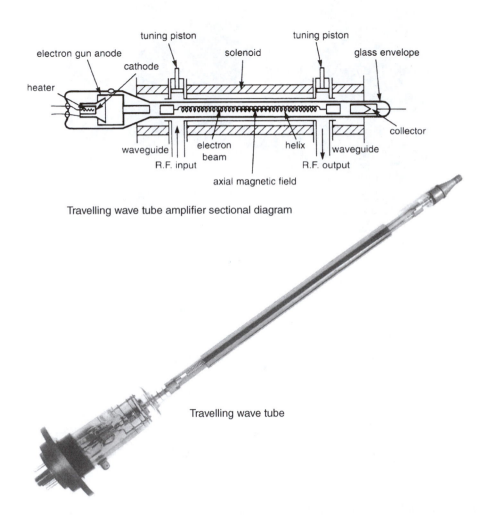

Travelling wave tube amplifier sectional diagram

Travelling wave tube

Figure 11.6   Travelling wave tube amplifier (Standard Telephones and Cables Ltd) output stage amplifier for microwave radio-relay systems

readily designed for operation at frequencies of up to at least 20 GHz without involving unduly critical mechanical tolerances. Designs for low input noise amplifiers and for high output power amplifiers became possible. Important applications in communication satellite systems and in military radio surveillance systems also followed.

The invention of the travelling-wave tube was due to Rudolf Kompfner, when working at the Clarendon Laboratory, Oxford during the war years – and it was through our joint membership of the wartime UK Co-ordination of Valve Development Committee that the writer became aware of the possibilities of the travelling-wave tube for radio-relay systems and sponsored the further development work by STC.

11.9 The invention of the travelling-wave tube

Rudolf Kompfner (1909–1977)

Rudolf Kompfner was born in Vienna in 1909 where he trained as an architect at the Technische Hochscule. He came to England in 1934 and practised as an architect for a time, but was interned as an alien at the outbreak of war in 1939. On his release he joined an Admiralty group working on valve research for radar at the University of Birmingham, under the direction of Professor L.M. Oliphant. It was in this group that Randall and Boot invented the multi-cavity pulsed magnetron that played such a vital role in the radar systems that helped the Allies to win the war.

Kompfner's own story of the invention of the travelling-wave tube is told, with characteristic modesty, in his book of the same title. His work with the Birmingham group had initially been concerned with attempts to reduce the noise in klystron amplifiers with a view to their use in the first stages of radar receivers. This work necessarily involved a study of the principles of velocity modulation of electron beams, first published by O and A. Heil in 1935. The aim was a low-noise amplifier that would improve the sensitivity of radar receivers, and thus increase the radar range, above that achievable by silicon-crystal frequency mixer first stages. [17]

However, the travelling-wave amplifier came about not directly from the klystron work, but rather from an attempt to devise a high-frequency oscilloscope by slowing down the signal wave so that it could interact with, and deflect, an electron beam. The slow-wave structure Kompfner chose was a wire helix – a stroke of genius because it was by no means apparent that a microwave would follow the turns of the helix. By November 1942 the possibility of using such a structure as an amplifier, but with a hollow electron beam surrounding the wire helix, had occurred to him and his notebook records: 'A completely untuned amplifier?' After many false starts, and an eventual realisation that the microwave field on the helix had an axial component that could interact with an axial electron beam and produce velocity-modulation 'bunching', success was achieved in November 1943 with a tube that produced a modest power gain of 1.5 times.

By 1944 a power gain of 10 times had been achieved in a travelling-wave tube with less than half the noise level of typical crystal mixers – and the principle of the travelling-wave tube had been fully and clearly demonstrated.

In 1944 Kompfner moved on to the Clarendon Laboratory, Oxford where he continued work on the travelling-wave tube, aiming at a theory that would enable design to be optimised and performance predicted. A visit to the Clarendon laboratory in that year by Dr J.R. Pierce of Bell Laboratories produced a partnership in the study of the travelling-wave tube that yielded a number of valuable results – notably Pierce's more precise theory that enabled gain and other performance characteristics to be accurately predicted. This theory revealed that some loss in the helix was actually beneficial in that it enabled higher gain to be achieved with stability; it also indicated that the useful bandwidth could exceed an octave.

The advantages of the travelling-wave tube were not lost sight of by the Bell Laboratories – after the TD2 4 GHz triode-based radio-relay system, the next higher frequency systems used travelling-wave tubes.

Rudolf Kompfner joined the Bell Laboratories at Holmdel in 1951 as a Director of Research, where his unique powers of scientific leadership were of great value to teams working on communication satellites, lasers and radio-astronomy. In 1973 he began a new career as a professor of engineering, sharing his time between Oxford and Stanford universities. He received many awards and distinctions, including Honorary Doctorates from the universities of Vienna and Oxford, and in 1976 the American President's Award for Achievement in Science – a remarkable record for someone who had no formal training in science and indeed did not embark on a scientific career until he was well into his thirties.

It was said of Rudolf Kompfner that '... he was a man of perfect integrity, great charm and boundless generosity' – the writer is proud to have known him as a friend and a war-time colleague.

11.10 The international standardisation of radio-relay systems

Beginning in the early 1950s, and following the lead given by the USA, the use of microwave radio-relay systems by the post, telephone and telegraph administrations throughout the world began to expand on a large scale – with distinctive contributions to the engineering of such systems by the UK, France, Germany and Japan. [5]

In order that radio-relay systems may provide satisfactory transmission of telephony, television and data over long distances, and to facilitate their interconnection with one another and with line systems, certain common technical characteristics are necessary. In particular the radio-frequencies to be used and the modulation characteristics of the radio carrier waves must be defined. These characteristics are specially important when links across international boundaries are involved.

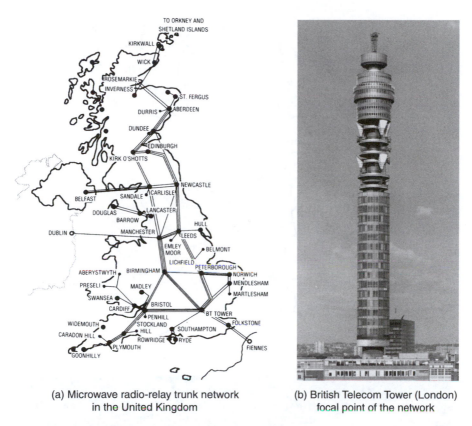

(a) Microwave radio-relay trunk network in the United Kingdom

(b) British Telecom Tower (London) focal point of the network

Figure 11.7 British Telecom (former British Post Office) microwave radio-relay trunk network (1980)

International agreement on preferred standards has been achieved through the work of the International Radio Consultative Committee of the International Telecommunication Union in Geneva, through its Study Group IX (Radio-Relay Systems). These standards have not only greatly facilitated the setting up of radio-relay links across national frontiers, they have also given a valuable lead to equipment manufacturers enabling them more efficiently to meet the needs of a growing world market. The standards are a remarkable example of the ability of engineers from a wide range of countries – including the USSR – to reach agreement on common solutions to technical problems, when these are based on scientific and engineering facts and agreed operational needs. [6]

It is satisfying to record that one of the first applications of the CCIR radio-relay system standards was to the 4 GHz multi-channel telephony and television link across the English Channel established by the British Post Office and the French PTT Administration in 1959, using equipment made by STC in England and LMT in France.

(a) Microwave radio-relay (TV) network in Europe (1961)

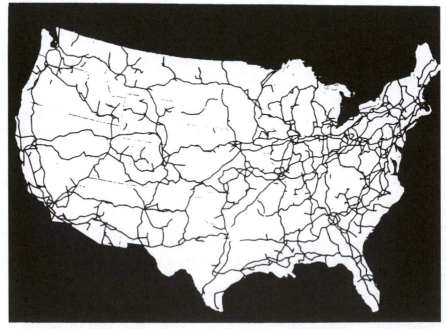

(b) Microwave radio-relay (TD2) network in USA (1980)

Figure 11.8   Growth of microwave radio-relay networks in Europe and USA

An interesting outcome from that pioneering experiment across the same path in 1931!

References

1   A.J. Gill, 'Chairman's Address, Wireless Section of the Institution of Electrical Engineers', (*Jl. IEE*, 84, No. 506, February 1939).
2   Gene O'Neill (ed.), *A History of Engineering and Science in the Bell System: Transmission Technology, 1925–1975*, Chapter 7 (ATT Co./Bell Laboratories, 1985).
3   J.H.H. Merriman and R.W. White, 'Frequency Modulation Tests on the Holyhead–Douglas Ultra-Short Wave Radio Link', (*Post Office Engineering Department Radio Report* No. 806, September 1942).
4   L. Lewin *et al.*, 'Phase Distortion in Feeders', (*Wireless Engineer*, 27, 1950).
5   Helmut Carl, *Radio-Relay Systems* (Macdonald and Co., London, 1966).
6   W.J. Bray, 'The Standardization of International Microwave Radio-Relay Systems' (*Proc. IEE*, 108 pt.B No. 38, 1961).
7   A.G. Clavier and L.C. Gallant, 'Anglo-French Micro-ray Link between Lympne and St. Inglevert' (*Electrical Communication*, 12, 1934).
8   R.H. and S.F. Varian, 'A High-Frequency Oscillator and Amplifier' (*Jl. Appl. Phys.*, 10, 1939).
9   E.M. Deloraine, 'Evolution de la Technique des Impulsions Applique au Telecommunications', (*Onde Electrique*, 33, 1954).
10 H.T. Friis, *Seventy-Five Years in an Exciting World*, (San Francisco Press, 1971).
11 H.T. Friis and A.C. Beck, 'Horn-Reflector Aerial', (U.S. Patent No. 2,236,728, filed Oct. 1936, issued March 1941).
12 G.N. Thayer et al., '*A Broad-band Microwave Relay System* between New York and Boston' (*Proc. IRE*, 37, 1949).
13 A.C. Dickieson, *The TD2 Story – from Research to Field Trial* (Bell Lab. Record, 45, October 1967).
14 W.J. Bray, 'Post Office Contributions to the Early History of the Development of Television in the UK' (*British Telecom Jl.*, 5 No.1, January 1987).
15 R.I. Clayton *et al.*, 'The London–Birmingham Television Radio-Relay Link', (*Proc. IEE*, 98 Pt.1, 1951).
16 G. Dawson and L.L. Hall, 'The Manchester–Kirk o'Shotts Television Radio-Relay System' (*Proc. IEE*, 101 Pt.1, 1954).
17 Rudolf Kompfner, *The Invention of the Traveling-Wave Tube*, (San Francisco Press, 1964 and *Wireless World* (UK), 52, November 1946).

Chapter 12

The inventors of the transistor and the microchip:
a world-wide revolution in electronics

12.1 Significance of the transistor and the microchip

The inventions of the transistor in the late 1940s and the planar integrated circuit – the microchip – in the 1960s began a development in electronics that was to have a wide-ranging, profound and continuing impact on telecommunications, sound and television broadcasting, and on computing throughout the world. They enabled electronic equipment to be made that was more compact, more reliable and lower in cost and power consumption than was possible using thermionic valves. The microchip in particular enabled circuit operations of far greater complexity to be performed reliably, rapidly and economically, greatly enhancing the capability of computers to calculate the service functions available in electronic telephone exchanges and the quality of colour television broadcasting.

The transistor and the microchip facilitated the design of larger capacity land and submarine cable systems and the design of communication satellites. They made possible a vast new range of customer equipment for computing, communicating and broadcasting; they gave rise to new electronic industries and changed old ones beyond recognition. [1] [2]

Transistors exploit the property of mono-crystals, in particular of the atoms of germanium and silicon (both in Group 4 of the periodic table), to conduct electricity by way of two separate carriers whose presence depends on the extent to which crystals have been made impure, at very low levels, by the addition of atoms of Group 3 (notably boron or indium) or Group 5 (notably phosphorus and arsenic). By suitable design active circuit elements such as rectifiers, amplifiers and switches can be created in small physical sizes, limited only by the amount of electrical power to be handled, and with long life expectation. On the other hand thermionic valves involving the emission of electrons from heated filaments or cathodes are necessarily bulky and, when mass produced, tend to be of limited life.

Initially transistors tended to be used individually on 'wired-circuit' boards, connected to other circuit elements such as resistors and capacitors by hand-made or pre-formed wiring involving operations by human beings – a process which set a limit to miniaturisation and also reliability. The next, and giant, leap forward was to integrate all the components on a single slice of material, usually silicon, giving rise to the planar integrated circuit or microchip.

This brilliant concept paved the way for almost unlimited miniaturisation – with many thousands, even millions, of transistors or other circuit elements on a few square millimetres of silicon. The microchip made possible highly auto-mated manufacturing processes, offering greatly enhanced reliability and cost reductions achieved by large-volume production.

12.2 The pre-history of the invention of the transistor

Solid-state devices in the form of crystals, e.g. of galena and carborundum, had been used in radio receivers for detection of modulated carrier waves from the beginning of the radio art, as had copper-oxide rectifiers for converting alternat-ing current to direct current. There were reports in the British and American technical press (*Wireless World*, May 1920 and *QST*, March 1920) describing a radio receiver designed by Dr E.W. Pickard, in which an oscillating crystal diode in a circuit containing a DC power source was used for heterodyne reception of continuous-wave carrier signals. The discovery of crystal diode oscillation has been attributed to Dr W.H. Eccles, some ten years earlier and 'crystals' had been used as mixers at the front end of radar receivers and microwave repeaters. But these were all essentially diodes. What was needed was a solid-state device capable of amplification. [3]

It has been said that 'had it not been for de Forest's Audion triode valve, the transistor would have been invented much earlier than 1948'. Be that as it may, there were signs from the 1920s onwards that the triode valve might not be the only possible device for generating and amplifying high-frequency oscillations.

In 1923 Dr Julius E. Lillienfeld (pioneer of the electrolytic capacitor) in Canada demonstrated a 'tube-less' radio receiver in which solid-state devices acted as oscillators and amplifiers. Between 1925 and 1928 Lillienfeld applied for Canadian patents for devices that had the geometry of today's insulated-gate field-effect transistors.

The German engineer O. Heil (inventor of the Heil velocity-modulation tube) filed a British patent in 1934 for a device based on a copper-oxide rectifier with an added electrode. In 1936 Holst and van Geel of the Philips Laboratories in Holland patented a type of bi-polar device; Van Geel continued the work on solid-state devices during the War years, filing other patents in 1943 and 1945. [4]

However, none of these efforts came to fruition, perhaps because of the lack of in-depth scientific knowledge of the phenomena involved, of detailed know-ledge of the materials technology and of the engineering skills needed to evolve practical designs. It remained for scientists and engineers of the Bell Telephone Laboratories in the USA to fill this gap.

12.3 The invention of the transistor

12.3.1 Three men who changed our world

This was the proud title that appeared on the front page of the Bell Laboratories Record in December 1972 [5], a commemorative issue describing events that led up to the invention of the transistor in 1948 and the developments that followed in the next 25 years. The three men, then research physicists in the Bell Laboratories and who jointly received the Nobel Prize in Physics in 1956, were:

- Dr John Bardeen;
- Dr Walter H. Brattain;
- Dr William Shockley.

John Bardeen, who joined Bell Labs in 1945, was co-inventor with Walter Brattain of the point-contact transistor. He received his early scientific education at the Universities of Wisconsin and Princeton, later becoming President of the American Physical Society and member of the US President's Science Advisory Committee. He received many honours, including a second Nobel Prize in 1972 for work on super-conductivity, and became Professor of Electrical Engineering and Physics at the University of Illinois.

Walter Brattain took MA and Ph.D. degrees at the universities of Oregon and Minnisota. He joined Bell Laboratories in 1929, working on the surface properties of solids and semi-conductor materials. His studies on the latter led him, jointly with John Bardeen, to the invention of the point-contact transistor. His many honorary awards included D.Sc. degrees from the universities of Portland and Minnesota; he was a member of the US National Defence Committee and the National Academy of Sciences.

William Shockley joined Bell Laboratories in 1936, having graduated at the California Institute of Technology and received a Ph.D. degree at the Massachusetts Institute of Technology. He is credited with the invention of the 'junction transistor' – a major step forward that overcame the limitations of the point-contact transistor. His war-time work included anti-submarine research and direction of a major solid-state research programme.

In 1955 he established the Shockley Semi-conductor Laboratory at Palo Alto, California and continued work in the transistor field until 1965. His awards included honorary degrees from the universities of Rutgers and Pennsylvania, and a Professorship of Engineering Science at Stanford University in 1963 which he held until 1975.

Shockley, who died in 1979, was a highly controversial figure in a field far from semi-conductor research – no less than control of the ongoing evolution of the human race by genetic manipulation, a theme which occupied him

THE KEY INVENTION

The work of early designers laid a foundation which was waiting for the key invention. The arrival of the key was first announced with this newspaper cutting from the New York Times in 1948. It was given only a small space on an inside page. It announced the discovery of the transistor. From this small announcement our story develops – slowly at first but now at great, and ever increasing, speed. The first transistors were laboratory curiosities quite unsuited to mass production. It was more than 10 years – about 1960 – before production techniques were developed which allowed photographic – like printing processes to be applied to the mass production of transiustors and then integrated circuits of greater and greater complexity. This process was called the PLANAR process because all of the manufacturing processes were carried out on one side of a flat wafer of

The original point contact transistor made at Bell Laboratories USA.

material. After this the speed of change increased as mass production led to falling prices and increasing reliability.

The first use of the PLANAR process was to make single transitors – each one on its own part of the silicon wafer.

Bardeen, Shockley and Brattain inventors of the transistor.

Figure 12.1   The invention of the transistor (1948)

(Bell Laboratories Record, December 1972)
John Bardeen, William Schockley and Walter Brattain

increasingly in the last two decades of his life. His proposals, which aimed at raising the average level of human intelligence by discouraging population increase amongst lower levels of intelligence, were found by many to be totally unacceptable. A proposal that he be awarded an honorary degree at Leeds University in 1973 was withdrawn. He was a singularly difficult person to deal with – he insisted on tape-recording virtually all interviews, and claimed that the 'tape-recorder was the single most important application of the transistor'. Nevertheless, his contributions to the development of the transistor were profound and far-reaching in their effects. [6]

12.3.2 The physics and technology of the transistor

To understand a little of what the inventors of the transistor – a term suggested by J.R. Pierce of Bell Laboratories – had achieved, an outline of the physics and technology involved may be helpful.[1]

The atoms of solids are bonded together by a combination of forces. For the elements silicon and germanium the four valency electrons (the outermost shell) play a key role. Each atom is sited at the centre of a regular tetrahedron formed by its four nearest neighbours, bound to each of them by two coupling electrons – one donated by it and the other by its neighbour.

There are no surplus, free, electrons in a pure crystal to conduct electricity when an electric field is applied. If however some of the atoms are replaced by atoms of elements in Group 5 (e.g. phosphorus or arsenic), their fifth valency electrons are surplus to bonding requirements and are free to conduct electricity; the material is then said to be 'n' (for negative)-type. On the other hand if the replacement is by atoms of elements in Group 3 (e.g. boron or indium), there are insufficient bonding elements and what are called 'holes' are created. These holes are also free to move in an electric field though with lower mobility than the electrons in n-type material and do so as if they carried a positive charge, when the material is said to be 'p' (for positive)-type.

If a piece of pure semi-conductor is doped in one region to be n-type and to be p-type elsewhere, the junction between the two will not conduct electricity when the n-region is made positive with respect to the p-region; the potential barrier at the junction cannot be overcome by the barriers on either side.

When the polarity is reversed a current will flow, and both the electrons and the holes crossing will have finite life-times in regions where normally they do not exist. In an n-p-n sandwich structure, with the first n-region (the emitter) negatively biased with respect to the p-region (the base) and the second region positively biased, we have the junction transistor proposed and analysed by Shockley, even before he and his colleagues had fabricated one.

The electrons injected into the base region diffuse with little loss to the second junction, which is favourably biased for their immediate collection. If a signal is applied to the low-impedance emitter-base terminals, that derived from the

---

[1] Contribution by Dr J.R. Tillman

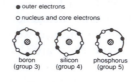

Atomic structures

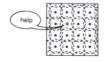

Phosphorous in silicon
(An n-type dopant)

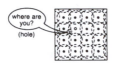

Boron in silicon
(A p-type dopant)

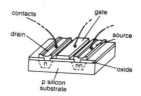

Schematic of an MOS transistor

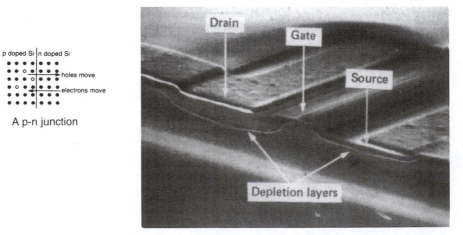

A p-n junction

Section through an MOS transistor

Figure 12.2   How an MOS junction transistor works

high-impedance collector-base terminals can show very considerable power amplification. A p-n-p sandwich will behave similarly, with electrodes biased in the opposite directions to those for n-p-n.

Bardeen and Brattain's point-contact transistor in fact used a thin slab of germanium with gold and tungsten contacts – reminiscent of the 'cat's whisker' crystal detectors in the early days of radio broadcasting. It demonstrated amplification but lacked stability and was not suited to large scale manufacture. Shockley's work suggested the replacement of the point contacts by two more layers of doped germanium with different doping from the centre slice – and the 'junction transistor' was born. Not only did it offer amplification in a remarkably compact and reproduceable device, it could function effectively with low-voltage DC input supplies much smaller than were required by the thermionic valve it replaced.

One of Shockley's earliest ideas had been to try to control the flow of electrons in a semi-conductor device by an electric field imposed from outside. Later in the 1950s this possibility of control by an external electric field was re-investigated and the result was the 'field-effect transistor' – the type most commonly used today. (For more detailed information, see R.G. Arns in Further Reading section.)

But equally important with the demonstration of experimental devices that worked and were repeatable, was the gradual creation of a viable physical theory that explained not only how transistors worked but which provided a basis for quantitative design and ongoing development.

12.3.3 And these were their words

In the commemorative issue of the Bell Lab Record of December 1972 – *25 Years of Transistors* – the President of the Laboratories and the inventors of the transistor made the following observations:

James B. Fisk (1834–1976)

The transistor began with basic reseach, in Bacon's words, into the 'secret motions of things' – the mysteries of solid-state physics has led to 'the enlarging of the bounds of human empire'. But the greatest impact of the transistor may be yet to come – if its benefits can serve as a reminder to society that solutions to many of our problems today, as well as human progress, will depend heavily on the continued pursuit of science and its responsible application.

And this was written before the microchip had revolutionised electronics and its applications in telecommunications, broadcasting and computing!

John Bardeen (1908–1991)

They were very exciting days after the invention of the point-contact transistor. To get a good patent it is necessary to have a good understanding of the basic mechanisms, and there were still questions about just how the holes flowed from the emitter to the collector. How important was the surface barrier layer in the transfer of holes from emitter to collector? Shockley initially suggested the junction transistor structure to help under-stand the mechanism. Independently, John Shire put emitter and collector points on opposite sides of a thin wafer of germanium, and found that the arrangement worked as a transistor. I can still remember the excitement I felt when I first learned of this dis-covery, which showed definitely that the holes in the emitter could flow appreciable distances through the bulk of n-type germanium.

This reference to John Shire's vital contribution is interesting and ought to be remembered when discussing the invention of the transistor!

Walter Brattain (1902–1987)
The transistor and the development of solid-state electronics were technical off-spring of the revolution in physics that was born with the conception of the principles $E = hv$ and $E = mc^2$ by Planck and Einstein at the beginning of this century. Since I am the oldest of the three Nobel Prize winners, I can say it all happened in my life-time.

The use of the transistor of which I am proudest is the small battery-operated radio. This has made it possible for even the most under-privileged peoples to listen ... as all peoples can now, within limits, listen to what they wish, independent of what dictatorial leaders might want them to hear and I feel that this will ultimately benefit human society.

William Shockley (1910–1989)
The creation of solid-state electronics was a complex effort. So many people contributed significantly that I would have to write a book to do them all justice. Instead I have focused on what might be over-looked – the foundation that supported the entire effort – the managerial skills that created the atmosphere of innovation at Bell Labs. These skills were essential to creating the transistor and to contributing the many benefits that have come to society as consequences.

12.3.4 Applications of transistors in telecommunications and broadcasting

During the two decades following their invention in 1948, at first slowly but with increasing tempo, transistors began to have a major impact on the design of telecommunication systems, improving performance and reliability, and reducing size and power requirements. Similar advantages became available in sound and television broadcasting equipment; broadcasting receivers, with mass markets measured in millions, benefited especially from the replacement of valves by transistors. Compact and efficient battery-operated receivers became available at relatively low cost.

The expanding role of transistors and other semi-conductor devices in tele-communications has been surveyed by Dr R. Tillman of the British Post Office Research Department; the following summary is based on that study. [2]

The junction transistor made its first application in telecommunications in transmission systems in the 1950s, then entirely analogue, by effectively replacing thermionic valves. By the 1960s the junction transistor had been developed to a point where inland inter-city cable systems carrying more than 10,000 frequency-division multiplex telephony channels on 9.5 mm diameter coaxial cables were possible, Chapter 10; transistors also replaced valves in the

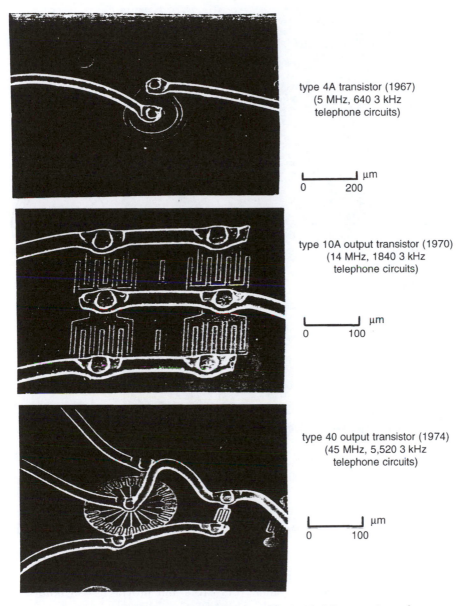

type 4A transistor (1967)
(5 MHz, 640 3 kHz
telephone circuits)

├──────┤ μm
0   200

type 10A output transistor (1970)
(14 MHz, 1840 3 kHz
telephone circuits)

├──────┤ μm
0   100

type 40 output transistor (1974)
(45 MHz, 5,520 3 kHz
telephone circuits)

├──────┤ μm
0   100

Figure 12.3   Post Office designed and manufactured high-reliability transistors for
submarine cable systems

(Since 1967 more than 10,000 transistors, valued at at least £5 million, have been made and delivered
for submarine system use by the Research Department at Dollis Hill and Martlesham).

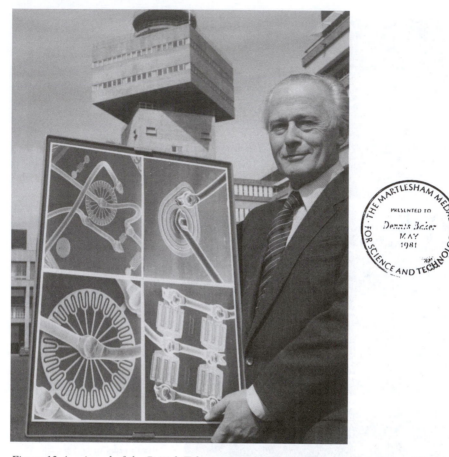

Figure 12.4   Award of the British Telecom Martlesham Medal to Dennis Baker (1981)
The award was made for his work on high-reliability transistors for submarine cable systems.

baseband and intermediate-frequency sections of microwave radio-relay systems (see Chapter 11).

Transistors found important applications in submarine cable systems, beginning in the mid-1960s. The UK was early in this field with several cables up to 1,500 km long, using silicon planar transistors and providing 480 both-way telephone circuits on a single cable. The ATT Co., using a 'germanium mesa transistor' developed by Bell Laboratories, laid an 800-circuit submarine cable between the USA and Spain in 1970.

Meanwhile scientists and engineers at the BPO Research Department had developed under the direction of Dr J.R. Tillman high-performance 'silicon transistors' of very high reliability which made possible a 1,800 telephone circuit submarine cable, designed and manufactured by Standard Telephones and Cables (UK) and laid between the UK and Canada in 1973/4. The reliability standards demanded by such systems with their many repeaters is extremely high and its achievement a demanding scientific and engineering task – typically there must be not more than one failure in 2,000 transistors over a period of 25 years.

Much credit for pioneering work in the design, assessment and manufacture of ultra-high reliability transistors for submarine cable systems is due to the BPO team responsible for this work: J.R. Tillman, D. Baker, and F.F. Roberts. They evolved methods for reliability assessment that included scientifically-based accelerated ageing tests involving very severe over-stressing by raising the operating temperature of transistors randomly selected from production batches, followed by electrical tests for any systematic or random changes in performance. The ability to carry out accelerated ageing tests with transistors, not possible with thermionic valves, was a significant factor in moving from valves to transistors in submarine systems. Their choice of silicon for submarine system transistors, as compared with the Bell Laboratories choice of germanium, was shown to be fully justified by the superior performance achieved with silicon. [7]

For its work on high-reliability transistors the BPO Research Department was awarded the Queen's Award to Industry in 1972. Dennis Baker was awarded the British Telecom Martlesham Medal in 1981 for his work in the semi-conductor field – a prestigious award made to present or former members of British Telecom staff who have made an outstanding contribution to telecommunications science and engineering. His reputation was world-wide – John Bardeen of the Bell Laboratories, co-inventor of the transistor, said 'it was a well deserved honour', and similar praise came from Ian Ross, President of the Bell Laboratories.

Although the first major applications of transistors in telecommunications were to analogue systems requiring linear amplification over a wide range of frequencies, the suitability of the transistor as a low power-consuming logic and switching element providing fast transition between clearly defined voltage or current levels, resulted in important new applications in computers, digital transmission (see Chapter 13) and electronic exchange switching systems (see Chapter 14). In association with other components, such as diodes and resistors, they formed the basis of logic gates enabling digital signals to be selected in an 'and' gate, combined in an 'or' gate, or rejected in a 'nor' gate). They also enabled 'bistable' circuits to be made.

In communication satellite systems, transistors and semi-conductor diodes provided low-noise microwave amplifiers, and silicon-based solar cells converting sun-light to electricity provided power for satellites (see Chapter 15).

But, above all, it was the reliability, small size, low power consumption, economy and versatility of transistors and other semi-conductor devices that provided the essential keys to major advances in the services that telecommunications, broadcasting and computers could provide.

Figure 12.5   The microchip revolution

Robert Noyce
(Intel)

Jack Kilby
(Texas Inst.)

Figure 12.6   The co-inventors of the planar integrated circuit – the microchip (1978/1979)

12.4 The next step forward in the electronic revolution

12.4.1 The invention of the planar integrated circuit – the microchip

Once it had been demonstrated that individual resistors and capacitors, as well as transistors and diodes, could be created on silicon by suitable doping with boron or phosphorus, it was perhaps inevitable that the idea should arise of forming and interconnecting a number of such components on one slice of silicon to make a 'planar' integrated circuit. (The term 'planar' indicates that the devices are formed in a single uniform plane.)

The need for the integrated circuit first became apparent as the circuitry of military rocket control systems grew increasingly complex and the lack of reliability of conventional printed circuit boards with individually assembled and wired electronic components became more evident – and, in some cases, disastrous.

From simple beginnings of just a few components formed on a single slice of silicon to make an oscillator or amplifier, a remarkable technology evolved that ultimately enabled the increasingly complex circuitry of computers, electronic exchange switching systems and television receivers to be built in finger-nail size microchips.

Early ideas for an integrated circuit were put forward by G.W.A. Dummer of the UK Royal Aircraft Establishment in 1952, with military control systems in mind, but these ideas were not followed up. It remained for two Americans: Jack S. Kilby and Robert N. Noyce to receive the accolade in October 1989 of the Draper Award and a prize of $175,000 each from the US National Academy of Engineering for 'their separate co-invention of the single-crystal integrated circuit, better known as the semi-conductor microchip'. The award was also based on 'their skill in bringing the integrated circuit to successful production and application in commercial products'. An illuminating pen-picture of both men and the steps that led them to the invention of the integrated circuit has been given by T.R. Reid. [8]

Jack Kilby was born in Jefferson City, Missouri in 1923; he graduated at the University of Illinois in 1947 and received a Master's degree in electrical engineering from the University of Wisconsin in 1950. He joined Texas Instruments (TI) and it was there that his work on the integrated circuit came to fruition in July 1958. He retired from TI in 1970 with more than 60 US patents to his credit. His many honours include the US National Medal of Science, and the US Institute of Electrical and Electronics Engineers Sarnoff and Zworykin medals.

Robert Noyce was born in Burlington, Iowa in 1927; he received a doctorate in physical electronics from the Massachusetts Institute of Technology in 1953. After leaving college he worked for a time in the Schockley Semi-conductor Laboratory. He became co-founder of Fairchild Semi-conductor Corporation in 1957, and in 1968 co-founded with G.E. Moore the Itel Corporation with the objective of making large-scale integration a practical reality. He holds 16 patents for semi-conductor devices and his many honours include the US

National Medal of Science in 1979, the US National Medal of Technology in 1987, and the UK Institution of Electrical Engineers Faraday Award.

At Texas Instruments Jack Kilby's first crude attempt at an integrated circuit used a slice of germanium on which were formed a transistor, a capacitor and three resistors to make a simple phase-shift oscillator. On 12 September 1958 the world's first integrated circuit worked and a new era of electronics was born. However, the components in Kilby's microchip were linked by fine gold wires, and a further important step remained to be taken – the deposition of connections as part of the manufacturing process, a step to which Noyce made a vital contribution.

At Fairchild Semi-conductors in January 1959 Robert Noyce wrote in his notebook:

... it would be desirable to make multiple devices on a single piece of silicon in order to make connections between the devices as part of the manufacturing process, and thus reduce size, weight etc as well as cost per active element.

And thus the idea of the monolithic integrated circuit was born.

Amongst their fellow engineers Kilby and Noyce are referred to as co-inventors of the microchip, a term that both men find satisfactory. However, it took several years of legal argument before Texas Instruments and Fairchild came to an agreement to cross-licence their patents, and make them available to other firms on a licencing basis.

12.5 The ongoing march of semi-conductor devices and technology

The years that followed the events of 1958/9 saw an intensive development of the design, technology and the manufacturing processes of the integrated circuit, leading to dramatic increases in the numbers of devices that can be accommodated on each chip and corresponding reductions in cost per device. The number of devices on each chip a few millimetres square has progressed from tens in the 1960s to millions or more in the 1980s, that is from small-scale, through medium-scale to large-scale integration, with the prospect of even larger-scale integration to come. This has led to microchips capable of increasingly diverse and complex functions, of high reliability at modest cost.

The manufacturing techniques evolved over the two decades since the invention of the transistor begin with the production of an ultra-pure silicon crystal in rod form which is then sliced into thin wafers. Some hundreds of chips or more are then formed on each wafer by refined photo-lithographic techniques in which light is projected on to a light-sensitive coating on the wafer through a mask bearing images of the desired chip patterns. The masks are themselves created from a single large drawing of a chip pattern, via a step and repeat camera and photo-reduction.

After chemical development of the chip patterns on the wafer, the doping of required areas and the formation of conducting connections by metallic

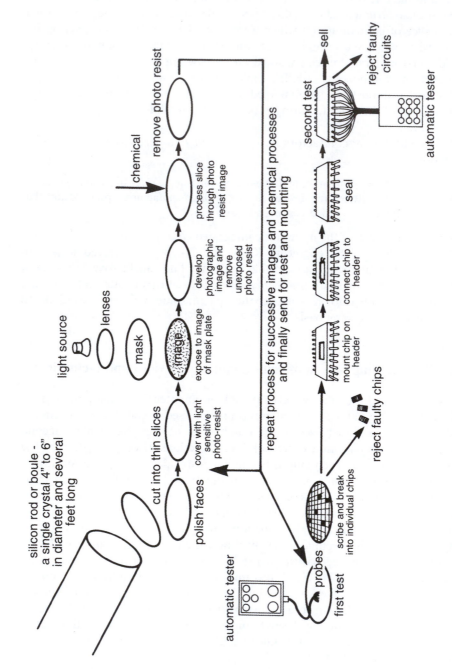

Figure 12.7 Basic manufacturing and testing process of the microchip

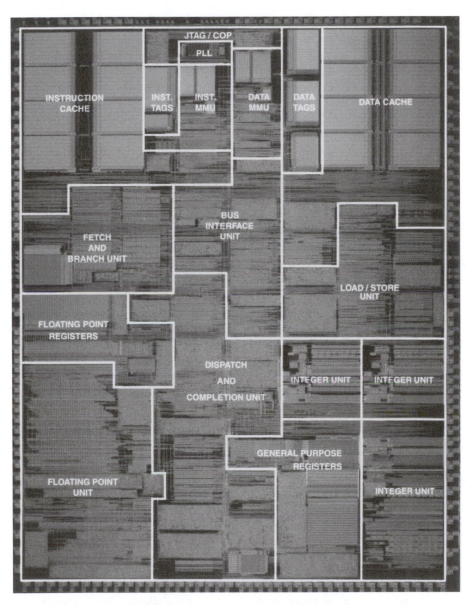

Figure 12.8   Large-scale integrated circuit (microchip) processor for personal computers and word-processors

Contains 3.6 million transistors on a 12.5 × 16 mm silicon chip with 0.65 micron spacing.

deposition is then carried out. Finally the chips are separated from the wafer, mounted on a header and sealed. These processes are carried out by highly automated manufacturing and testing equipment, under conditions of extreme cleanliness. It is by automation of the manufacturing processes and careful control to produce a high yield of tested and proven chips from each wafer that the cost per chip is kept down. [1] [9]

These developments have made possible memory devices capable of holding in digital form the equivalent of hundreds of thousands of words, micro-processors that control the functions of a computer or other electronic control process, e.g in an electronic telephone exchange switching system (see Chapter 14) and even, in the limit, a complete computer on a single chip.

And beyond these levels of integration, still larger-scale integration using electron-beams or X-rays instead of visible light, is possible, reducing the device size to molecular proportions, with many millions of devices per chip. At present the microchip exists only in planar, i.e. two-dimensional, form. It is not inconceivable that it could become three-dimensional, with a further vast increase in capacity – perhaps enabling hour-long television programmes to be stored on a single chip.

From the first step in semi-conductor technology that began in 1948 many other devices have evolved, some for use at microwave frequencies such as the IMPATT diode, charge-coupled devices (CCDs) and light-emitting diodes (LEDs) that have found application in television cameras and visual displays, and solid-state lasers for optical fibre transmission systems. The year 1948 was indeed the beginning of a major revolution in electronics that transformed telecommunications, broadcasting and computing.

Germanium, the semi-conductor first used to make transistors on any large scale in the 1950s, and by Kilby for the first integrated circuit, soon gave way to silicon for several reasons. But other semi-conductors, based on compounds of elements of Group 3 with those of Group 5 of the Periodic Table, e.g. gallium-arsenide and indium-phosphorus, first explored by H. Welker of Siemens in the 1950s, have become important for devices such as lasers, light-emitting and IMPATT diodes.

However, not all inventors of integrated circuit technology accept the priority of Noyce and Kilby. A press report in the UK *Sunday Times* in 1990 [10] states that a 52-year-old computer expert Gilbert Hyatt living in Los Angeles, California had been granted a patent for the invention of a 'blueprint' that would enable a computer circuit to be etched on a silicon chip and create the basis for a micro-processor. Hyatt's patent was first registered in 1968, and it has taken 20 years of litigation and expert study of counter claims before the patent was granted. The report also claims that the first commercial micro-processor, made by Ted Hoff at the Intel Corporation, USA in 1971, was based on concepts originated by Hyatt.

References

1 W.H. Mayall, *The Challenge of the Chip* (Science Museum, H.M. Stationery Office, 1980).
2 J.R. Tillman, 'The Expanding Role of Semi-conductor Devices in Tele-communications', (*The Radio and Electronic Engineer*, 43 No.1/2, Jan/Feb 1973).
3 H.E. Stockman (USA), 'Oscillating Crystals' (letter in *Wireless World*, May 1981).
4 W. Gosling. 'The Pre-History of the Transistor' (*Radio and Electronic Engineer*, 43 No.1/2, Jan/Feb 1973).
5 *25 Years of Transistors* (Bell Laboratories Record, December 1972).
6 Obituary Notice, Prof. William Schockley (*Daily Telegraph*, 15 August, 1989).
7 M.F. Holmes, 'Active Element in Submerged Repeaters; the First Quarter Century' (*Proc. IEE*, 123 No.10R, October 1976).
8 T.R. Reid, *The Chip: The Micro-Electronics Revolution and the Men Who Made It* (Simon and Schuster, 1985).
9 *The Micro-chip Revolution* (British Telecom Education Service, ES57, 1985).
10 'Father of the Micro-chip On Line for Millions at Last', (*Sunday Times*, 9 September 1990).

Further Reading

P.R. Morris, *A History of the World Semiconductor Industry* (IEE History of Technology Series, No.12, 1990).
E.H. Cooke-Yarborough, 'Some Early Transistor Applications in the UK' (*IEE Education and Science Journal*, June 1998).
Robert G. Arns, 'The other transistor – early history of the metal-oxide-semiconductor field-effect transistor' (*IEE Science and Education Journal*, October 1998).

Chapter 13

The creators of information theory, pulse-code modulation and digital techniques

13.1 Information theory as a vital key to progress

The better understanding of the nature of information, e.g. as conveyed from a sender to a receiver by telephonic speech, a facsimile picture or a television programme, and the development of theories providing a quantitative approach to design have been vital for progress in the evolution of more efficient telecommunication transmission and broadcasting systems. The initial, and major, steps forward were the work of a few innovative and far-sighted individuals in the the laboratories of the ATT Co.mpany in the USA, Standard Telephones and Cables Ltd in England and the International Telephone and Telegraph Company in France.

The existence of the sidebands of a modulated carrier wave had been demonstrated mathematically by J.R. Carson and G.A. Campbell of the ATT Co. Bell Laboratories in 1915 (see Chapter 6); their work was extended in the 1920s and 1940s to include the concept of information theory by two other members of the laboratories R.V.L. Hartley and C.E. Shannon. [1]

R.V.L. Hartley
R.V. Hartley was a Rhodes Scholar who had graduated from the University of Utah and later carried out original work on valve circuits – including the invention of an oscillator circuit that bears his name – in ATT's Western Electric Laboratories. His interest in information theory was stimulated by earlier work by Nyquist of ATT Co. on the maximum rate at which telegraph signals could be transmitted in a given frequency bandwidth.

His studies, published in 1928, resulted in rules for the design of transmission circuits that were in use for more than two decades until C.E. Shannon greatly broadened their scope, and introduced the effects of noise in limiting the rate at which a communication channel could transmit information. [2]

C.E. Shannon: master of information theory (dec. 2001)

Claude E. Shannon has been described as a mathematician and educator whose information theory profoundly changed scientific perspectives on human communication and greatly facilitated the development of new technology for the transmission and processing of information. It became an important milestone in the transition of society from an industrial base to an information base. His *Mathematical Theory of Communication* was first published as a Bell System Monograph in 1948, and later in book form with Warren Weaver. [3]

He was born in 1916 at Gaylord, Michigan and graduated in electrical engineering at the University of Michigan, later receiving MS and Ph.D. degrees in Mathematics from the Massachusetts Insitute of Technology. He joined Bell Laboratories as research mathematician in 1941 and in 1956 became Professor of Communication at MIT where he held the Donner Chair of Science. He received many honours, including the US National Medal of Science and an IEEE Prize Award.

Shannon's Master's thesis, completed at the age of 21, established important relationships between Boolean algebra and the design of switching networks and computers, that is still regarded as a classic in its field. He also made significant contributions to the theory of cryptography as applied to communication security. His early work on cryptography, issued in 1945 as a confidential monograph *Communication Theory of Secrecy Systems*, discussed variable coding transformations that made the encyphered text seem random. Some of these ideas emerged later in his information theory which centred on the idea that coding was essential to all communication, i.e. the process of encoding a message into a form suitable for transmission and decoding the message by applying the inverse of the encoding transformation. The performance of a communication channel in the presence of noise could be assessed mathematically from appropriate statistical descriptions of the message information and the noise.

The full development of Shannon's information theory is complex and involves abstruse philosophical concepts as well as sophisticated mathematical analysis. However, one result in particular is of direct value to communication engineers. It involves the concept that the smallest unit of information is the 'bit', represented by a simple 'on–off' signal. This enables the information-handling capacity C of a communication channel with a signal-to-noise power ratio S/N in a bandwidth B to be given by

$$C = B \log (1 + S/N) \text{ bit/sec.}$$

The application of this formula reveals that most practical communication systems have considerable redundancy. For example a telephone channel of 3 kHz bandwidth and 40 dB (10,000:1) signal-to-noise ratio has a theoretical information-handling capacity of 40,000 bits/sec. Whilst some practical systems operate with rates that are a fraction of this value, it is to be noted that many present-day modems operate at 56 kbit/s. [3] [4] [5]

However the result does give a target against which the efficiency of a given coding system may be measured – an important consideration for example in television transmission where the redundancy of practical systems is generally very high, i.e. the efficiency is very low, and the theoretical scope for improvement is considerable.

Claude Shannon died on 24 February 2001, aged 84. Sadly, his last years were marred by Alzheimer's disease and he was too ill to attend a ceremony arranged by the American Institute of Electrical and Electronic Engineers, the Information Theory Society and the University of Michigan where a statue to his honour was unveiled by his wife.

The Times obituary (12 March 2001) records:

Eccentric and at times even erratic, Claude Shannon single-handedly laid down the general rules of modern information theory, creating the mathematical foundations for a technical revolution. Without his clarity of thought and sustained ability to work his way through intractable problems, such advances as e-mail and the World Wide Web would not have been possible.

13.2 The digital revolution

A 'digital' system is one in which information, whether in the form of speech, facsimile, television or data signals, is transmitted or processed by 'on' and 'off' pulses transmitted by modulation of an electric current, radio or light carrier wave, as compared with an 'analogue' system in which the continuously varying amplitude of the information signal is conveyed by corresponding amplitude, frequency or phase modulation of a carrier wave. Not all digital systems are 'binary', 'ternary' and other multi-level codes are also used in line and radio systems.

Digital techniques, which first began to find application in the 1930s, have now revolutionised telecommunications, broadcasting and computing by providing superior quality of transmission and recording, greater flexibility in the processing of information, and substantial economic advantages relative to analogue systems. In the case of terrestrial or satellite transmitters using the radio-frequency spectrum digital techniques enable more transmissions/channels to be accommodated in a given bandwidth of the spectrum.

These advantages derive from the basic 'on–off' simplicity of digital signals which enables noise and interference to be rejected, non-linear distortion avoided, and a number of communication channels interleaved in time on a common transmission path, e.g. on cable, waveguide or radio, without mutual interference.

The principle of time-interleaving was recognised as early as the 1850s in the Baudot multiplex telegraph system which used a mechanical commutator to interleave a number of telegraph channels on a single pair of wires. An attempt was made by Miner in 1903 similarly to interleave speech channels by sampling the speech waveforms with a mechanical commutator at 4,000 samples a second – before information theory indicated that good quality required a sampling rate

at least twice the highest speech frequency. It is also relevant that Morse's telegraph, see Chapter 3, was essentially a coded digital system.

In the USA P.M. Rainey of ATT Co./Bell suggested in 1926 that the
elements of a picture might be represented, for transmission purposes, by coded
binary, i.e. 'on–off', signals to circumvent imperfections in the transmission
medium (USA Patent No.1,608,527, November 1926).

In France pulse-time multiplex methods were studied from 1932 at the laboratories of Le Matériel Téléphonique, where Deloraine and Reeves were granted
patent rights on pulse-time and pulse-code modulation methods in 1938. [6]
Pulse-time and pulse-length modulation are not 'digital', they are examples of
'analogue pulse modulation'.

During the war years the British and American Armies used 8-channel pulse-
time modulation microwave radio-relay systems, but civil applications lagged
until the 1960s when the principles and advantages of pulse-code modulation
became better understood and the technology for implementation – notably the
transistor and microchip – became available.

Until the 1960s frequency–division multiplexing on cables and frequency-
modulation radio-relay systems were the dominant modes long distance
communication (see Chapters 6 and 11).

13.3 Pulse-code modulation: the key invention

Until the advent of pulse-code modulation, digital communication systems had
used width or time modulation of individual pulses, each bearing a single voice
or other channel. Such systems achieved little, if any, signal-to-noise advantage
compared with frequency modulation systems using the same mean transmitted
power. The key invention, that conferred on digital systems a virtually complete
immunity to noise, interference and non-linear distortion, was pulse-code modulation – an invention that created a revolution in the transmission, switching and
storage of information.

Pulse-code modulation involves sampling the instantaneous amplitude of any
analogue signal, whether speech, facsimile or television, at an appropriate rate
and translating each sample into a short train of pulses or 'code', with a number
of pulses in each train determined by the type of signal and the quality of
transmission required. For example, telephone speech is typically sampled at
8,000 times per second and each sample is coded into 8 pulses, corresponding to
2 to the power of 8, i.e. 256, discrete amplitude levels of the speech signal
waveform, the pulse rate for a single speech channel then being 64,000 bits per
second. On the other hand a high-quality television signal may require a pulse
rate of 100 Mbit/sec.

However, more sophisticated coding techniques – embodying some of the
concepts foreshadowed by Shannon on the nature of information – can reduce
the pulse rate required for television signals substantially.

The noise immunity of digital PCM signals is created by the ability to re-generate the signals, e.g. in repeaters, by circuits that respond only to each pulse amplitude above a predetermined level and reject noise below that level.

The coding of each signal level sample into a group of on-off pulses (called a 'byte') is usually carried out in 'binary notation'. In binary notation a decimal number, which could represent the level of a signal sample, is represented by a succession of on or off pulses in which the first stands for one, the second for two, the third for four, the fourth for eight and so on. Thus a sample of level decimal number 11 is represented by binary 1101, i.e. one plus two, no four, and plus eight.

However, digital pulse-code modulation involves a substantial increase in the frequency bandwidth required for transmission; for example a 64,000 bit/sec telephone speech circuit requires a bandwidth, according to Hartley's law, of at least 32 kHz, compared with the basic speech bandwidth of about 3 kHz. This, together with the lack of suitable device technology, delayed the practical application of digital PCM for two decades or more after its invention in 1937 when wide bandwidth transmission systems became available.

The benefits of pulse-code modulation may be summarised as follows:

• digitally encoded signals may be transmitted over long distances, requiring many repeaters in tandem, or switched in many exchanges, without introducing noise, distortion or loss of signal strength, whereas analogue signals would suffer a progressive loss of quality;
• many different types of signal, e.g. speech, data, facsimile and television, can be accommodated on a path such as a pair of wires, a coaxial or optical fibre cable, or a microwave carrier, without mutual interference;
• digital signals may be stored, e.g. on magnetic tape, computer discs or solid-state devices , without loss of quality due to repeated use or other causes;
• digital PCM signals lend themselves to the use of microchips for coding, de-coding, regeneration, and information storage, with their advantages of low-cost, reliability and small size.

Alec Harley Reeves (1902–1971): Inventor of PCM

Alec Reeves was born in Redhill, Surrey, England in 1902. After graduating in electrical engineering at the City and Guilds Engineering College, London he joined the staff of International Western Electric (later part of Standard Telephones and Cables Ltd) and worked on long-wave radio communication across the Atlantic in the 1920s. Over the next ten years he contributed to the development of short-wave and microwave radio systems. His work in this area, and especially on multiplex telephony, undoubtedly made him aware of the need of a more effective means for reducing noise and crosstalk in such systems.

In his lifetime Reeves was awarded more than 100 patents, but by far the most important and far-reaching in its effect was that of pulse-code modulation made in 1937 when he was working in the Paris laboratories of Le Matériel Téléphonique. His British PCM patent was granted in 1939.

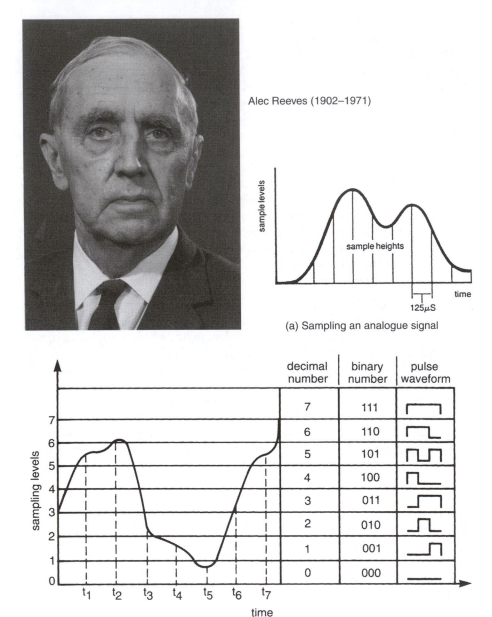

Alec Reeves (1902–1971)

(a) Sampling an analogue signal

(b) Coding an analogue signal into a binary pulse waveform

Figure 13.1   Alec Reeves and the invention of pulse-code modulation (1937)

Following the outbreak of the Second World War, Reeves and other colleagues escaped from Paris in June 1940 to work in the laboratories of Standard Telephones and Cables Ltd in England. By nature a peace-loving man Reeves was reluctant to work on offensive weapons and his war-time work concentrated on precision bombing aids – including OBOE – which, since it could define military targets with remarkable accuracy, would in the end save civilian lives. For this contribution to the war effort he was awarded the Order of the British Empire.

He was in charge of research on electronic switching systems and devices at Standard Telecommunication Laboratories until his retirement at the age of 69, but continued work in his own research company until his death two years later. This included studies in optical communications on contract to the British Post Office (now British Telecom). It is a tribute to the character of Alec Reeves that only a few days before his death from cancer he was discussing with the writer, then Director of Research of the BPO, his forward-looking ideas for direct amplification of light in an optical fibre communication system.

A modest and retiring man, Reeves spent much of his private life helping others, including youth and community work and assisting in the rehabilitation of those who had served prison sentences. His rewards came late in his life when the value of his work on pulse-code modulation came to be understood. In 1969 he was awarded the honour of Commander of the British Empire – and the unique distinction of the issue of a British Post Office stamp commemorating the invention of PCM! The University of Surrey, with the support of Standard Telecommunications and Cables, has created a Chair of telecommunication studies in his name. He was also awarded the Columbus Medal and the Franklin Medal, and an honorary Doctorate of the University of Essex.

In 1990 the Institution of Electrical Engineers, London created an annual award to be known as the 'Alec Reeves Premium Award' to be given each year for the best paper on digital coding as judged by the appropriate Professional Group of the Institution – a fitting and continuing tribute to a remarkable man. A colleague of Alec Reeves, Professor K. Cattermole of the University of Essex, has written an illuminating account of the life and work of a man who was one of most outstanding and innovative scientists of his day [7].

To quote Professor Cattermole on Alec Reeves:

He expressed his belief in a better era with 'a much saner set of values, based on truth so far as seen, combining science with direct intuitive knowledge' [*note his implicit faith that such a combination is possible*].

Alec Reeves would never have set himself up as an ideal. But his personal integrity, his combination of hope and rationality, of inventiveness and ethics, sets as fine an example as I have ever seen.

13.4 The applications of pulse-code modulation

Perhaps the earliest application of PCM, albeit on a small scale, was in a valve-operated speech security system developed in Bell Labs during the Second World War for use on short-wave radio links.

Any large-scale use of PCM had to await the availability of reliable, mass-produced transistors, which began in the early 1960s. The first large-scale application was made in the USA by the Bell system on wire-pair cables providing multi-channel voice circuits between telephone exchanges ('junction circuits'), commencing in 1962 [8]. This system enabled more channels to be provided on existing cables, with better transmission quality and lower cost. The system – T1 carrier – provided 24 voice channels on a 1.5 Mbit/sec digital carrier with 8 kHz sampling of a 4 kHz voice band and 8-bit logarithmic encoding. Seven bits of the code were used for voice and the eighth for signalling, i.e. for setting up and supervising exchange connections. This in-built

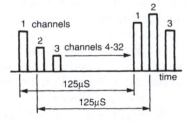

(a) Time-division multiplex of 32 telephone channels
(with amplitude modulation of channel pulse samples)

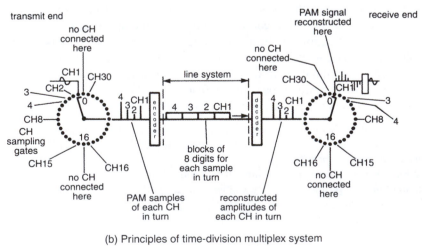

(b) Principles of time-division multiplex system
using pulse-code modulation

Figure 13.2   Time-division multiplexing

signalling feature proved valuable in improving the over-all economics of the digital system, as compared with analogue systems using out-of-band signalling.

The overall economics were further improved when, in the 1970s, it became possible to integrate digital transmission on wire-pair junction cables with the first Bell System 'electronic' trunk exchange ESS1. However, the first ESS electronic exchanges were essentially analogue space-division systems. Digital transmission on inter-city coaxial cables had difficulty at first in competing economically with the well-established analogue frequency-division multiplex cable systems. Digital microwave radio systems bearing 720 or 1440 voice channels at 45 or 90 Mbit/sec nevertheless proved successful for short inter-exchange links.

In the UK the advantages and potential of PCM were recognised early and studies initiated in the 1950s at the Post Office Research Laboratories, Dollis Hill to optimise the parameters of PCM systems, e.g. sampling frequency, pulses per sample, coding law, signalling methods and synchronisation of pulse trains generated in different parts of the network for voice, television, facsimile and data signals. In the telephone transmission field R.O. Carter and Dr D.L. Richards made notable contributions, as did W.E. Duerdoth in the digital switching field, Chapter 14.

Although the use of asymmetric-sideband amplitude modulation for television broadcasting had become universal – largely for reasons of bandwidth limitation in the available radio spectrum – the potential of digital techniques for inter-city television transmission, the conversion of line-standards, e.g.from US 525 to Europe 625 lines, the recording of television signals and the processing of television signals for bandwidth reduction, has been both recognised and implemented. The optimisation of standards for digital television transmission was also the subject of detailed study at the British Post Office Research Laboratories, in co-operation with the Research Department of the BBC and UK Industry – in this work Dr N.W.J. Lewis, I.F. Macdiarmid and J.W. Allnatt at Dollis Hill made important contributions.

These studies, carried out in co-operation with Member Countries and Operating Agencies, led to the recognition by the International Telephone and Telegraph, and Radio Consultative Committees of the International Telecommunication Union of a 'hierarchy' of digital PCM systems in steps ranging from 30 to 7680 telephone channels at pulse bit rates ranging from 2 Mbit/sec to 565 Mbit/sec. with 'slots' for data, facsimile and television channels. This international co-operation was of great importance in facilitating communication across national boundaries, and in enabling the development of equipment to be carried out in an orderly and efficient manner.

In the UK the first practical implementation of digital PCM was in 24-channel (later 30-channel), 2 Mbit/s systems on wire-pair inter-exchange junction cables in the middle 1960s, leading to a world first PCM digital exchange at Empress, London in 1968 (see Chapter 14).

Digital PCM microwave (11 GHz) radio systems providing 2,000 voice channels, or a television channel, on each of six carriers at 140 Mbit/sec were also developed and used in the British Post Office trunk network.

A multi-disciplinary task force set up by the BPO in the 1970s to study modernisation of the UK telecommunications network made a clear recommendation, on economic and operational grounds, to 'go digital' for the main inter-city trunk network, both for transmission and switching, and to plan for an 'integrated digital network' in which the benefits of digital working could be extended to customers premises for a wide variety of telecommunication services.

The advent of optical fibre cable with its very wide useful bandwidth, and solid-state lasers that could be readily pulse modulated, opened another door to the deployment of digital PCM techniques (see Chapter 17). Satellite communication systems use digital techniques because they offer more effective use of the satellite microwave power, give greater flexibility in the use of the satellite communication bandwidth and enable the signals from different Earth stations to be readily interleaved in the satellite transponder, Chapter 15.

References

1 *A History of Engineering and Science in the Bell System: The Early Years (1875–1925)* (Chapter 2, Research in Communication Theory, p. 909, ATT - o./Bell Laboratories, 1975).

2 R.V.L. Hartley, 'Transmission of Information' (*Bell System Technical Jl.*, July 1928).

3 C.E. Shannon and W. Weaver, *The Mathematical Theory of Communications* (Urbana University, Illinois Press, 1949).

4 J.R. Pierce, 'The Early Days of Information Theory' (*IEEE Transactions on Information Theory*, 19.1, 3–8, 1973).

5 E.C. Cherry, 'A History of Information Theory' (*Proc. IEE*, London, 98.3: September 1951).

6 E.M. Deloraine, 'Evolution de la Technique des Impulsions Applique au Telecommunications' (*Onde Electrique*, 33, 1954).

7 K.W. Cattermole, 'The Life and Work of A.H. Reeves', (presented at the Institution of Electrical Engineers, London Discussion Meeting on *Milestones in Telecommunications*, November 1989); K.W. Cattermole, 'A.H. Reeves: The Man Behind the Engineer' (*IEE Review*, November, 1990).

8 E.F. O'Neill (ed.), *A History of Engineering and Science in the Bell System: Transmission Technology (1925–1975), part III, The Advent of Digital Systems* (ATT Co./Bell Laboratories, 1985).

Chapter 14

The pioneers of electro-mechanical and computer-controlled electronic exchange switching systems

14.1 Exchange switching systems: the essential link between you and the world

With more than 1,000 million telephones in use in the world and with facsimile, data and video communication growing at an ever increasing rate, the problem remaining once transmission paths have been created by cable, radio or satellite is to enable each customer to find and communicate with another anywhere in the world. Today this problem has been solved through the efforts of a number of pioneering scientists, engineers and mathematicians and their successors who have evolved efficient, economic and reliable exchange switching systems to provide the necessary inter-connection between customers, whether in the same town or on opposite sides of the world.

The creation of a world-wide telecommunication network has necessarily involved international co-operation on a large scale to secure a commonality of technical and operational procedures to ensure that both local and long-distance connections can be made with equal ease – and which has been achieved through the International Telephone and Telegraph Consultative Committee (CCITT) of the International Telecommunication Union. The world telecommunication network may well be claimed to be the most extensive, complex, costly and indispensable artefact yet created by man – it is in fact the 'nervous system' of civilised mankind. The contribution of exchange switching engineers to the creation of this artefact has been critical, demanding both innovative skills and intellectual ability of a high order.

14.2 The electro-mechanical era of exchange switching

From the 1900s to the 1970s the automatic exchange switching systems of the world were electro-mechanical, that is they used electrically powered relays, stepping mechanisms or motors to establish connections between an incoming voice channel and a desired outgoing channel selected from banks of tens or hundreds.

The history of switching technology is at once complex, detailed and uses a language all its own; the following brief summary is based mainly on the book by Robert J. Chapuis, the former Head of the Switching and Operations Department of the CCITT in Geneva, [1] the *Early History of Engineering and Science in the Bell System*, [2] the *Story of the Telephone in the UK* [3] and papers presented at an Institution of Electrical Engineers (London) meeting on *Milestones in Telecommunications* [4]. A comprehensive and up-to-date account of the principles and practice of signalling is available in R. Manterfield's book *Telecommunication Signalling* (1999) [17].

The first automatic electro-mechanical system was the Strowger step-by-step system (described in Chapter 4), invented in 1889 and which in various modified forms dominated the world's exchanges for nearly a century. The Strowger step system went through a number of stages of evolution at the hands of several inventors, notably by Keith and the Erickson brothers, and Smith and Campbell in the USA from about 1896 to 1915, with the aim of handling economically, with as few switches as possible, the ever growing numbers of customer lines at each exchange.

The basic concepts that made step-by-step a viable switching system were:

(a) the ten point rotary dial which generated pulses of line current according to the digit dialled;
(b) the two-direction 100-point switch bank which could be controlled by the dialled pulses;
(c) the use of successive switching stages, connected by transfer trunks via 100-point switches;
(d) the line switch which enabled a given customer's line to be connected to a first selector switch via a free path in a trunk multiple;
(e) the line finder which looked for an energised line in a multiple of lines and connected it to a trunk and first selector (i.e. the line finder was thus the inverse of the line switch).

It was the application of these concepts that contributed substantially to the economic viability and the quality of service, i.e. reduction of calls failed due to non-availablity of plant, of automatic exchange switching systems throughout the world during the first half of the 20th century, and the principles of which, with different technology, found applications in the electronic switching systems that followed.

14.2.1 Trunking and grading

An innovation that significantly improved the efficiency of switch utilisation was the 'graded–multiple' principle invented by E.A. Gray of ATT Co. in 1905. It involved a departure from the strictly decimal organisation of the network of

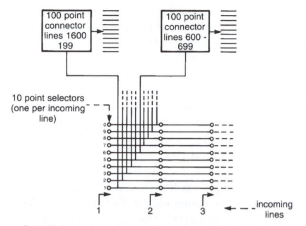

(i) 1,000-line system using transfer trunks and 10-point selectors

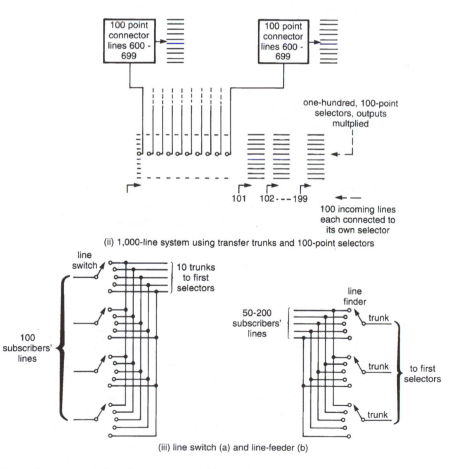

(ii) 1,000-line system using transfer trunks and 100-point selectors

(iii) line switch (a) and line-feeder (b)

Figure 14.1   Principles of exchange switching systems

switches in an exchange by allocating, say, five trunks as first choices feeding perhaps seven selectors, so that fewer selectors are required to handle a given volume of traffic.

14.2.2   Traffic engineering

Developments such as this inspired a new concept – 'traffic engineering' – which seeks answers to questions such as 'how many calls are to be expected over a given pathway at a given time and what quantity of plant is required to handle these with a defined low probability of failure?'. It involves statistical descriptions of traffic flow, the notion of a 'busy hour' in the day when calls are most numerous, and the use of probability theory to assess the percentage of failed calls likely with a given plant configuration. G.T. Blood and M.C. Rorty of ATT Co. made early contributions to traffic theory but major steps forward, made in the early 1900s and which remained in almost continuous use since then, were made by E.C. Molina:

E.C. Molina (ATT Co.): pioneer of telephony

Molina was an extraordinary individual who left a lasting mark on the evolution of telephony. He joined Western Electric at the age of 21 with no formal education beyond high school and proved to be a versatile and prolific inventor, making important contributions to machine switching. He joined the Research Department of ATT Co. in 1901, and during his long career in this company and with Bell Laboratories, contributed extensively to the development and application of traffic mathematics. This was made possible by a disciplined self-study of mathematics that led him to become an expert in certain fields, particularly in the work of Laplace. In recognition of this work, and the encouragement it gave to young people for self-study, he received in 1952 the Honorary Degree of Doctor of Science from Newark College of Engineering, an institution at which he taught mathematics until 1964, some 20 years after his retirement from the Bell System. [3]

Traffic engineering studies had also received attention in Europe, notably in Denmark by A.K. Erlang.

A.K. Erlang: pioneer of traffic theory

Erlang initiated a new approach to traffic theory by introducing an exponential distribution of call holding time, compared with the constant average holding time assumed in earlier studies. This enabled greater accuracy to be achieved in the prediction of call failures, and permitted analytic solutions to traffic flow problems not previously possible; it also provided a basis for modern traffic 'queuing' theory. However, for a time Erlang's work received little attention outside his own country, and was virtually ignored in the USA. Its publication in the *British Post Office Electrical Engineers Journal* (London) in 1918 brought to general notice a theory of major importance in the design of switching systems that was eventually widely used.

14.2.3 Exchange switch development

Although the 100-point (10 by 10) point Strowger switch retained a first place in the world's exchange switching systems for several decades after 1900, a variety of other types of switch were developed in the USA and Europe, stimulated by the need to keep down the cost of exchange equipment as the numbers of customer lines at each exchange increased into the tens and hundreds of thousands. Typical of these were the linear 'Panel' switch used by the ATT Co. in America, which simplified trunk multiple wiring, and the 'Rotary' switch used in France and elsewhere in Europe. Switch points increased to 200 or more and electric motor drives were used to power the switches in place of direct control by dialled pulses of current from the customer lines.

'Crossbar' switches, pioneered in Sweden and which used plane arrays of rectangularly co-ordinated contacts, were also developed and widely used.

The problems of switch point contact reliability were tackled in Europe and the USA, for example, by 'Reed Relay' exchanges in which the contacts were contained in glass tubes sealed against the atmosphere and operated by externally applied magnetic fields.

However, it became increasingly clear that switch development alone did not adequately cope with the problems created by telephone traffic growth – especially evident in big cities such as New York and London. A new approach was required to deal with this situation.

14.2.4 The 'big city' problem: evolution of common control of exchange switching

The heart of the 'Big City' problem was that the volume of exchange switching equipment using step-by-step direct control – despite switch development – increased more rapidly than the number of customer lines so that a point was reached where such systems became uneconomic and unwieldy beyond a few thousand lines. The solution lay in what came to be known as 'Common Control', a principle which was later applied in computer-controlled electronic exchanges as well as electro-mechanical ones.

In a common-control switching system the pulses created by the dialling of a call are placed in a temporary store or register and then converted into a machine language best adapted to control the interconnecting switches. The stored signals, in this converted form and under the direction of a built-in program, then control the switches needed to make the desired connection, initiate the process of alerting the called customer and informing the calling customer of the progress of the call.

Common control provides greater flexibility in the routing of calls and the handling other types of traffic as well as telephone calls, and the control of fault conditions, by appropriate design of the stored program. It embodies many of the features of computers, e.g. in the use of memory, translation, stored-program and machine-language, all of which were later achieved in micro-electronic, as compared with electro- mechanical, format.

14.2.5 The London Director System

Col. Purves and G.F. Odell

Following the end of the First World War the British Post Office initiated the progressive introduction of automatic telephone exchanges throughout the country, based on the Strowger step system. It was early recognised that London presented a 'Big City' problem in common with that encountered in the USA. In 1919 consultations took place between BPO staff headed by Col. Purves, then Engineer-in-Chief, and ATT Co./Western Electric which led to a decision in 1922 to develop a London Director system of automatic telephony based on the common-control register/translator principles, using Strowger step switches and incorporating the combined letter/numeral numbering system first proposed by W.G. Blauvert of ATT Co. [2] [3].

The first London Director System installation was made in 1927 – it was then described by G.F. Odell, BPO Asst. Engineer-in Chief, as a 'high water-mark in the tide of human creative intelligence' – one wonders how he might have described the vastly more complex and sophisticated computer-controlled electronic switching systems that came into being half a century later.

14.3 The computer-controlled electronic exchange switching era

The evolution of exchange switching from electro-mechanical systems to computer-controlled electronic systems opened the door to major improvements in the speed and reliability of the connection process, but above all it enabled additional service facilities such as abbreviated dialling, call transfer and item-ised billing of call charges to be provided rapidly and efficiently. It greatly facili-tated the management of switching exchanges and customer lines, especially with regard to fault location and correction. And it provided the flexibility needed to handle the expanding non-voice traffic such as Telex, data, facsimile and video.

This evolution was made possible mainly by the following advances in device technology and system concepts:

- the invention of the transistor and the microchip (see Chapter 12);
- the principle of 'stored-program control', adapted from computer systems;
- the use of digital techniques (Chapter 13).

The transistor and the microchip, with other solid-state devices, provided the basis for fast and very reliable switches, in contrast with slow moving electro-mechanical switches which required much maintenance. These devices enabled temporary and semi-permanent memories to be made for storing inform-ation, e.g. in transient form about call progress and, in more permanent form, the programs needed for switch-control. They also enabled 'logic' operations, e.g. the manipulation of information into a different form suitable for switch control, to be carried out extremely rapidly. By contrast with earlier 'wired logic', 'solid-state logic' could be readily re-arranged to modify or change the stored program to accommodate changing or new service requirements.

Many of the first electronic exchanges used 'space division' techniques, i.e. each switch handled one call at a time. With the increasing use of digital transmission in junction and trunk networks (see Chapter 13) it became desirable and economic to switch on a 'time-division' basis, each switch handling many calls interleaved in time – made possible by the fast operating speeds possible with solid-state devices.

Progress from electro-mechanical to computer-controlled electronic exchanges proved to be a complex process, involving massive development effort and a – sometimes painful – 'learning process' in which early experimental designs had to be subjected to the rigours of field trials on a sufficient scale to reveal any limitations.

Furthermore, technology was advancing more rapidly than development could be completed, notably in the following areas:

• the change from 'reed' to digital switching;
• the progress of storage technology from magnetic cores to ever more complex semi-conductor stores;
• the evolution of reliable software;
• the need for ever larger scale integration of semi-conductor devices on chips to solve difficult design problems cost-effectively, especially analogue to digital conversion on a per-customer basis.

The following account draws mainly on experience with early electronic exchange design and field trials in the British Post Office and the UK telecommunications industry, and in the Bell System in the USA, beginning in the 1960s. Because of the complexity of the work involved these developments were very much team efforts, and any names mentioned in this short account as 'pioneers' are essentially those of some of the team leaders. It has also to be remembered that similar developments were taking place in other countries, including France, Germany, Sweden and Japan. [5]

14.4 Electronic exchange development in the UK

Following the end of the Second World War the British Post Office Research Department at Dollis Hill, London, recognising the potential advantages of electronic switching, began experimental studies concentrating mainly on a multiplexed approach with time-division pulse-amplitude modulation (PAM-TDM) of the speech path as the most likely to yield an economic solution. Major activities in the mid-1950s included the design of electronic register-translators for the London Director and similar switching systems, and magnetic-drum translators for the Subscriber Trunk Dialling (STD) system.

These developments benefited significantly from experience gained during the war years in the design of the Colossus digital computer by BPO Research staff for cracking German military codes – probably the first full-scale digital computer ever made. [6]

The developments in electronic switching systems technology in the UK during the period from 1950 to 1975 are surveyed in depth in Hughes and Longden; [7] only the main highlights are described below.

14.4.1   The Highgate Wood field trial

The BPO and five leading UK manufacturers of exchange equipment agreed in 1956 to pool research effort in electronic switching in a Joint Electronic Research Committee (JERC). After examining various space- and time-division possibilities, and taking into account the available technology, it was decided to proceed with the design of a PAM-TDM system for an exchange capable of serving up to 20,000 lines. The electronic devices available then included early transistors, diodes, valves and cold-cathode tubes as switches; magnetic drums were available for code translation and recording class of service, call and line status information.

The field trial at Highgate Wood, NW London was commissioned in 1962 with 600 lines; however, by mid-1963 it was concluded that it did not provide the basis of a viable and economic design and was withdrawn from service. The main problems that arose were centred on excessive power consumption, high cost of components, inadequate noise and cross-talk performance and incompatibility of signalling with the existing network.

In retrospect the Highgate Wood field trial can be seen to have been over-ambitious, bearing in mind the limitations of device technology at the time. But many valuable lessons were learned and the TDM approach was referred back to research, later to emerge in the form of pulse-code modulation PCM-TDM switching systems.

It is significant that the Bell System also went through a 'learning experience' with its first electronic exchange at Morris, Illinois, when little of the technology used survived in later designs.

Meanwhile, the BPO and the manufacturers concentrated on a range of space-division common-control exchanges (TXE2 and TXE4) using reed-relay switches and well-tried transistor-transistor logic circuitry. These were not strictly electronic exchanges but their use on a large scale allowed time for the development of PCM-TDM switching and System X described below.

Experience with early models of reed-relay exchanges showed that the reed-relays, although sealed against deterioration of the contacts by the external atmosphere, nevertheless were not immune to contact failures and required great care in manufacture and application to ensure high reliability. The objective of wholly electronic exchanges remained and was pursued with vigour.

*Figure 14.2   The Right Hon. John Stonehouse, Postmaster General opening the
PCM–digital exchange at Empress, London with the author (1968)*

14.4.2 The first digital switching exchange: Empress, London (1968)

The success of pulse-code modulation time-division (PCM-TDM) transmission
over junction and trunk routes (see Chapter 13) pointed to the desirability, for
economic and operational reasons, for switching in tandem and trunk exchanges
without de-multiplexing to voice frequencies. On the other hand, the relatively
high cost of 'codecs' for coding voice signals into PCM format and de-coding,
prohibited the use PCM-TDM for local exchanges until such time as more
cost-effective device technology – e.g. the microchip became available.

Design studies of a PCM-TDM switching system suitable for a field trial
began at the BPO Research Department, Dollis Hill, London in 1965. The
trial system switched traffic on six 24-channel PCM links between three
neighbouring London exchanges.

Figure 14.3   PCM–digital telephone exchange at Empress, London

Opened in 1968 it was the first operational digital exchange to go into public service in the world.

It used a space-time-space switching configuration, with diode-transistor logic and some early designs of TTL integrated circuits. The switch paths were controlled by a cyclically addressed ferrite-core store. [8] [9]

The BPO field trial system was installed at Empress Exchange, West London and was opened for live traffic by the then Postmaster General Mr John Stonehouse in September 1968. It was followed by PCM-TDM tandem switching system, similar to Empress but with processor control of switching, built by Standard Telephones and Cables Ltd and installed at Moorgate Exchange, London in 1971.

The success of these early digital electronic exchanges led to an intensive study of the main UK network by a BPO Trunk Task Force which concluded in 1971 that:

integrated digital switching and transmission could offer significant economic advantages, which would increase as large-scale integrated circuits became available.

The experience gained with the Empress and Moorgate digital exchanges, and with the evolutionary modular design of the TXE2 and TXE4 exchanges, provided basic inputs to the next major step forward in exchange switching system design – System X.

14.4.3 The evolutionary approach to exchange switching: System X

The evolutionary System X concept, initiated by the British Post Office and subsequently developed in collaboration with UK industry, defined a common framework for a family of exchange switching systems including local, tandem and trunk exchanges, based on a series of hardware and software sub-systems each carrying out or controlling specific functions in the switching process. By so doing provision was made for future changes, e.g. to take advantage of new system concepts and advances in device technology, and to provide new services rapidly and economically, with a minimum of equipment replacement. At the same time, System X was designed to work with the existing network on either an 'overlay' or 'replacement' basis so that the network could be progressively up-dated without disruption.

Central to the System X design were:

- stored program control based on data processing and computer techniques providing a powerful and flexible means for controlling the operation of the exchange and performing administrative functions such as maintenance, recording call charges and collecting traffic data;
- common channel signalling which uses one out of perhaps hundreds of traffic channels for providing all the signalling functions needed to control the setting up and routing of calls;
- provision for integrated digital transmission and switching via local as well as tandem and trunk exchanges;
- the use of semi-conductor technology, including microchips, e.g. large-scale integrated circuits, as processors for the common-control of switching;
- provision for non-voice services such as facsimile, data, Prestel, and audio/video conference facilities.

Many of the System X concepts were initiated in the Research Department of the British Post Office at Dollis Hill, London and were later given specific shape by a Joint PO–UK Industry Advisory Group on System Definition (AGSD) set up in 1968 to define the architecture of System X and its sub-systems.

The subsequent collaborative programme for the design and manufacture of System X, involving Post Office Telecommunications (later British Telecom), General Electric Telecommunications Ltd, Plessey Telecommunications and Standard Telephones and Cables Ltd (until 1982), was initiated in 1976, with the

(a) System X – exchange unit

(b) System X – digital sub-unit

Figure 14.4   System X exchange and typical sub-unit

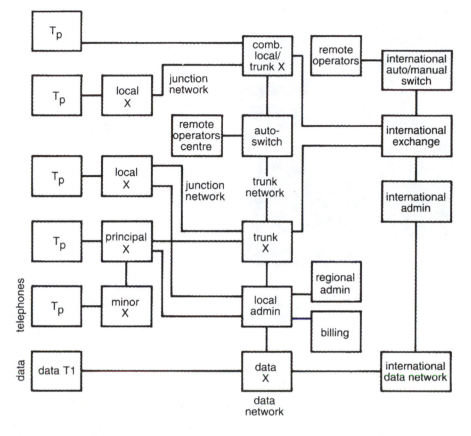

Figure 14.5   Types of System X exchanges

(Providing local, trunk and international exchanges, auto-manual and data-switching)

aim of large-scale production for the UK and world markets. Valuable contributions from UK industry were made by Jim Warman, who master-minded the reed-relay (TXE 1 and 3) developments, George Baker who pioneered the magnetic drum and Martin Ward for his work on stored program control.

14.4.4 British Post Office/telecom pioneers in electronic switching

Any development as complex and technologically challenging as the evolution of electronic switching inevitably involves many creative minds and many contributors to the ultimate outcome. The following have made particularly noteworthy contributions.

Dr Thomas Flowers

Dr Flowers was responsible for much of the early work in the Research Department of the Post Office at Dollis Hill, NW London on time-division electronic

Dr T. Flowers (1980)

C. J. Hughes (1984)

L. R. F. Harris and J. Martin (1983)

Figure 14.6   British Telecom Martlesham Medal winners for innovative work in electronic switching

switching that led to the Highgate Wood field trial. He is well known for his work during the war years on 'Colossus' – the world first digital electronic computer – which achieved major success in breaking German military communication codes. A number of the ideas inherent in Colossus, e.g. stored-program control and memory systems, later found application in electronic switching.

He was the first to be awarded the Martlesham Medal, in 1980, in recognition of his achievements.

L.R.F. Harris
Roy Harris and John Martin (see below) were jointly awarded the Martlesham Medal in 1983 for their work in master-minding for British Telecom the design and development of System X – it may fairly be said that Harris invented System X! He is the son of a former Director of Research and Engineer-in-Chief of the Post Office and had worked with Dr Flowers on early electronic switching research at Dollis Hill. He led the team which developed the early versions of the reed-relay stored-program commmon control exchanges that later became the TXE2 and TXE 4 exchanges widely used in the UK.

He made a vital contribution to System X with the concept of modular equipment, set within an enduring overall system 'architecture', with the functions of the modules or sub-systems, and the interfaces between them, precisely defined.

Roy Harris was born in 1926; he graduated with first-class honours in Mechanical Sciences at Cambridge University and joined the Post Office in 1947. He became Director of Telecommunications Systems Strategy in 1974 and has played a leading role in national and European forums aiming at harmonising development in telecommunications and information technology. There are some 20 patents bearing his name and he has published extensively on switching strategy.

The development of a ground-breaking project as complex as System X – with its involvement not only in new technology and system concepts but also in relationships between the BPO and UK telecommunications manufacturing industry – presented a challenge of the first order to all concerned. It was a challenge to the people involved – their personalities and loyalties to their employers, and the priorities of the new concepts they were creating. Roy Harris has given a fascinating 'insider view' of this story in his IEE Millennium Symposium Contribution on *Electronic Telephone Exchanges in the UK – Research and Development – 1960–1968*. [16] [17] (See also 'Electronic Telephone Exchanges in the United Kingdom: Early Research and Development: 1947–1963.)

John Martin
John Martin arrived in the System X team at the start of collaborative development with UK industry in the 1970s at a time when software-controlled digital micro-electronic technology was in its infancy. He became Director of System X Development and was responsible for overall management and leadership of the large joint PO–UK industry teams concerned with the development of System X.

He was born in 1927 and joined the Post Office Engineering Department in 1943 as a Youth-in-Training. He obtained a degree by part-time study at Northampton Polytechnic (London). His early work was on the introduction of subscriber trunk dialling in the UK and the development of reed-relay electronic exchange switching systems (TXE 3 and TXE 4).

In presenting the Martlesham Medal to Roy Harris and John Martin, John Alvey, then British Telecom Managing Director (Development and Procurement), said:

The successful development and supply of System X is the most significant technical event in British Telecom's recent history. The project is the backbone of our digital modernisation; on it ride all the new services which will help to transform the way in which Britain's commercial community goes about it's daily tasks.

Charles Hughes

Charles Hughes was awarded the Martlesham Medal in 1984. He pioneered the use for telecommunications what the world now knows as the 'micro-processor' – a vital component in the design of stored-program control switching systems.

His early work at the PO Research Department, Dollis Hill was in the management of the Empress Exchange digital exchange field trial and to which, in collaboration with a colleague Winston Duerdoth, he contributed important ideas in the use of PCM-TDM techniques.

Charles Hughes' achievements include:

- the micro-processor concept;
- establishing important principles in the architecture of stored program-controlled switching systems;
- contributions to the development of digital switching and recorded voice systems;
- the concept of a universal digital system of variable capacity accommodating instantaneous demands from a number of constantly varying sources.

Charles Hughes was born in 1925; he graduated from London University and received a Master's Degree. He joined the PO as an executive engineer in 1955 after working for a time with Cable and Wireless Ltd, working initially on radio research and then for many years on electronic switching. He became BPO Deputy Director (Switching and Trunk Transmission) in 1975, and in 1987 was awarded a Fellowship of Engineering (Royal Academy of Engineering, London).

14.5 Electronic exchange development in the Bell System (USA)

To the scientists, mathematicians and engineers of the Bell Laboratories must be awarded the accolade of pioneers of electronic exchange switching based on stored-program common control – their work in the early 1960s and onwards was followed with great interest by telephone engineers throughout the world. It stimulated similar switching developments in many countries, notably in the UK, Sweden, Germany, France and Japan. [5]

A lively, perceptive and entertaining account of switching developments in Sweden appears in John Meurling's book *A Switch in Time*. [10]

The proving of major advances such as stored-program control demands a field trial on a sufficient scale to demonstrate both practicability and economic viability – with the Bell System such a field trial took place at Morris, Illinois, from 1960 to 1962. [11]

14.5.1 The Morris, Illinois electronic exchange field trial

The design of the experimental exchange at Morris, with up to 400 customer lines, was based on the use of electronic rather than electro-mechanical techniques to control and carry out switching operations, enabling millions of individual functions to be carried out in the 10 seconds or so taken to dial a seven-digit number.

It was essentially a 'space-division' system using some 20,000 gas-filled tubes as switches, together with customer line scanners and a stored-program common control using a 'flying-spot' storage tube and photographic plates to provide a long-term memory. An electro-static 'barrier-grid' storage tube provided a short-term memory for dialled number and call progress information.

It is of interest that the use of gas-tubes as switches with their limited current-handling capability made it necessary to use tone-frequency signalling – this is also true of other electronic systems. Other advanced features in the trial were abbreviated code dialling and call transfer facilities.

Although little of the technology used in the Morris field trial survived in later electronic exchange development, the value of stored-program control, with its demonstrated operational flexibility, was fully proven and adopted in the systems that followed.

The faster setting up of calls available with 'touch-tone' dialling was appreciated by customers and created an incentive to wider use of this facility.

The experience gained from the Morris field trial, as with the British Post Office Highgate Wood and Empress field trials, provided both a stimulus and a basis for further electronic exchange development – notably in the Bell System with their 'No.1 ESS'. A detailed account of this development is given in a special issue of the Bell Laboratories Record for June, 1965, on which the following summary is based. [12]

14.5.2 The Bell System No. 1 ESS: a pioneering and major advance in electronic switching systems

The opening of the No.1 Electronic Switching Exchange at Succasunna, New Jersey in May 1965 has been described by W. Keister as 'the culmination of the largest single development project ever undertaken by the Bell Laboratories for the Bell System'. [12]

The basic concept of No.1 ESS is of a single high-speed electronic central processor operating with a stored program to control the switching operations

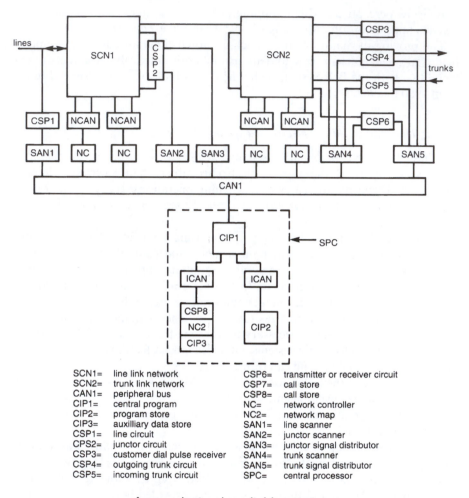

SCN1= line link network
SCN2= trunk link network
CAN1= peripheral bus
CIP1= central program
CIP2= program store
CIP3= auxilliary data store
CSP1= line circuit
CPS2= junctor circuit
CSP3= customer dial pulse receiver
CSP4= outgoing trunk circuit
CSP5= incoming trunk circuit

CSP6= transmitter or receiver circuit
CSP7= call store
CSP8= call store
NC= network controller
NC2= network map
SAN1= line scanner
SAN2= junctor scanner
SAN3= junctor signal distributor
SAN4= trunk scanner
SAN5= trunk signal distributor
SPC= central processor

A new electronic switching system

The evolution of telephone switching systems is toward common control. This first commercial electronic system is centered about a stored-program real-time data processor that facilitates modifications and additions to future service at low cost

Figure 14.7   The first electronic switching exchange Bell System (USA) (1965)

on a time-sharing basis. Through the switching network, inter-connections can be made between customer lines and trunks, and access provided to services needed to handle calls such as tones, signalling detectors and ringing. The central processor includes temporary and semi-permanent memories, the former for storing transient information required in the processing of calls dialled by the customer, while the latter stores the program for control of the switching operations.

Input and output information for the central processor is provided by scanners, e.g. of calling customer lines, and distributors, e.g. of signals initiating action required by the central processor.

Although wired logic is used in the central processor, changes in the program can be readily made in the program memory to accomodate a variety of services.

The device technology used in ESS 1 included:

- the 'ferrod', a magnetic current sensing device associated with each customer line;
- the 'ferreed', a two-state magnetic reed switch sealed in a glass envelope;
- the 'twistor' semi-permanent memory, a magnetic sheet capable of storing several million bits of information in magnetised dot form, the pattern of stored information being alterable by replacing a card between the magnetic memory planes.

Thus the ESS 1 system made effective use of a variety of magnetic device technology, in addition to silicon transistors in its logic circuits. In this sense it was perhaps less 'electronic' than its successors a decade or so later which used large-scale integrated semi-conductor circuits to carry out similar functions even more efficiently, economically and compactly.

Figure 14.8   R. W. Ketchledge, Director of Bell Electronic Switching Lab. (1959) (Author of From Morris to Succassuna ESS 1)

In reviewing progress in 1969 during an interview in *ESS – Four Years On* [13] R.W. Ketchledge, then Executive Director of Bell Laboratories Electronic Switching Division, noted that by that date more than 80 ESS systems were in regular service in the Bell System, and observed that:

The important thing about ESS is stored-program control . . . we have enough experience now to know we were absolutely right in 1955 when we gambled on this – it was a big gamble then – and turned the entire development around from a wired-logic system to a stored program system. We could see that stored program could give us a kind of flexibility that switching never had before by changing programs we can add new services more quickly and at lower cost than would be possible by changing hardware.

14.5.3 Bell pioneers and the onward march of progress in electronic switching

It is perhaps invidious to try to single out 'pioneers and innovators' in an operation as complex and challenging as the early development of electronic switching in the Bell System – so many people were involved not only in the evolution of new system concepts and the creation of new technology, but also in the innovative management of a huge project, with its involvement of manufacturing on a massive scale by Western Electric and field installation and testing throughout the USA.

An early pre-cursor of ESS 1 is worthy of mention for its pioneering content – the field trial of a stored-program, digital trunk exchange (ESSEX), developed by Earl Vaughan of BT and used in commercial service. He was the first to propose integrated digital switching and transmission, which he first demonstrated in 1959.

Some of the contributors to the ESS1 development as outlined above are:

- W. Keister: the evolution of telephone switching;
- R.W. Ketchledge: from Morris to Succassuna;
- J.J. Yostpile: features and service;
- E.H. Siegel and Jr S. Silber: the stored program;
- A.H. Doblmaier: the control unit;
- R.H. Meinken and L.W. Stammerjohn: memory devices;
- A. Feiner: the switching network. [12]

From the pioneering development of ESS1 other developments in electronic switching followed in the Bell System, including ESS2 and ESS3 for the smaller exchanges in suburban and rural communities. [5]

These, like ESS1, were essentially space-division switching systems. An early exploitation of the possibilities of time-division switching was the 101 ESS offered in 1963 as a private branch exchange for business users. [14]

By the 1970s PCM-digital transmission was beginning to supplement FDM-analogue transmission over junction and trunk routes in the USA as in the UK and it became operationally and economically desirable to switch in this mode. The outcome in the Bell System was the ESS4 PCM-digital electronic switching system, designed as a successor to electro-mechanical cross-bar

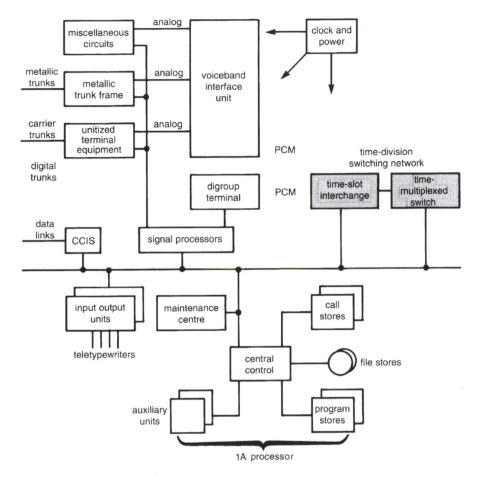

Figure 14.9   Bell System (USA) No.4 electronic switching system

The No.4 ESS routes incoming digital signals to the appropriate out-going trunks under the control of a local processor. Incoming analogue signals on wire or carrier circuits are first converted to PCM digital format.

switching systems, to cater economically and efficiently with the massive growth in cross-country telecomunication traffic then beginning. [15]

Perhaps Alec Reeves, the UK inventor of pulse-code modulation, would have been gratified to see, in the field of electronic switching, this outcome of his brain-child!

It is perhaps relevant to note however that progress towards ever more complex, highly-centralised control of nation-wide switching systems, dependent on computer control involving sophisticated software, is not without its problems. Difficult to detect software errors can – and have exceptionally – caused large-scale network 'shutdown' in the USA and elsewhere.

References

1 Robert J. Chappuis, *100 Years of Telephone Switching: 1878–1978* (North-Holland Publishing Co., 1982).
2 *A History of Engineering and Science in the Bell System: The Early Years (1875–1925)* (ATT Co. – Bell Laboratories, 1975).
3 J.H. Robertson, *The Story of the Telephone* (Sir Isaac Pitman and Sons (London), 1947).
4 Papers presented at The Institution of Electrical Engineers (London), Science, Education and Technology Meeting on: *Milestones in Telecommunications*, November 1989:
    (a) A. Emmerson, *Strowger and the Invention of the Automatic Telephone Exchange (1879–1912)*;
    (b) S. Welch, *The Early Years in the UK Telephone Service Environment*;
    (c) W. Duerdoth, *The Introduction of Digital Switching in the 1960s*;
5 Amos E. Joel Jr. (ed.), *Electronic Switching: Central Office Systems of the World* (Institute of Electrical and Electronics Engineers, New York, 1976).
6 A.W.M. Coombes, 'Colossus and the History of Computing' (*Post Office Elecl. Eng. Jl.*, 70, July 1977).
7 C.J. Hughes and T.E. Longden, 'Developments in Switching System Technology' (*Post Office Elecl. Eng. Jl.*, 74 Pt 3, October 1981).
8 K.J. Chapman and C.J. Hughes, 'A Field Trial of an Experimental Pulse-code Modulation Tandem Exchange', (*Post Office Elecl. Eng. Jl.*, 61, October 1968).
9 E. Walker and W.T. Duerdoth, 'Trunking and Traffic Principles of a PCM Exchange', (*Proc. Inst. Elecl. Eng.* (London), 111, December 1964).
10 J. Meurling and R. Jeans, *A Switch in Time* (Telephony Publishing Corp. USA, 1985).
11 E.G. Neumiller, *The Electronic Central Office at Morris, Illinois* (Bell Lab. Record, December, 1962).
12 *The No.1 Electronic Switching System* (Bell Lab. Record, June 1965).
13 R.W. Ketchledge, *ESS – Four Years On* (Bell Lab. Record, August 1969).
14 *101 ESS – A More Flexible Telephone Service for Business* (Bell Lab. Record, February 1963).
15 C.D. Johnson, *No.4 ESS – Long-Distance Switching for the Future* (Bell Lab. Record, Sept. 1973).
16 L.R.F. Harris, *Electronic Telephone Exchanges in the UK*, (IEE Millennium Symposium, History of Electrical Engineering, University of Durham, July 2000).
17 L.R.F. Harris, *Electronic Telephone Exchanges in the UK: Early Research and Development: 1947–1963*, Inst. Brit. Telecom Engineers Journal, Vol. 2., Pt. 4, Oct.–Dec. 2001.

Chapter 15

The first satellite communication engineers

15.1 The impact of satellites on world-wide telecommunications and broadcasting

The ability of Earth-orbiting satellites to provide, economically and reliably, direct high-quality communication between any two locations on the earth's surface, and direct broadcasting coverage over large areas, has expanded enormously the scope and volume of these services.

The technical advantages were obvious enough. For telecommunication links only one repeater was required, i.e. in the satellite, compared with the many repeaters on long land and submarine cables, and land microwave radio-relay systems, with substantial gains in reliability. For direct broadcasting to homes large areas could be covered by a single transmitter in the satellite, without encountering the 'shadow' areas created by hills and mountains when using ground-based transmitters.

Furthermore, the microwave frequency band most suited to satellite communication systems was immune to the vagaries of the ionosphere and was virtually free from atmospheric absorption over a range from about 1,000 to 10,000 MHz, above which rain absorption became progressively more apparent. By sharing frequency usage with compatible terrestrial microwave systems, wide frequency bands could be made available for satellite system use. This factor, coupled with the use of highly directional aerials at Earth stations, meant that there was scope for a massive growth of telecommunication traffic on a world-wide scale, involving Earth stations ranging from small units providing some tens of telephone circuits to large units with several thousands. Transmission over satellite microwave links proved to be virtually free from the fading and reflection phenomena encountered at times on terrestrial microwave links, and was suitable for high-definition television signals.

Long-distance telecommunication links by satellite also offered operational advantages in that these could leap-frog intervening countries, avoiding the troublesome problems of wayleaves and transit charges that terrestrial links were prone to. Because of the relatively high initial costs involved, notably for launching satellites, the economics of satellite links compared with land and submarine links, are more favourable for the longer distances. And for direct television broadcasting the initial cost per viewer by satellite improves, compared with ground-based transmitters, as the coverage and the viewing audience increases.

But until the middle 1950s the possibility of communication by satellite was considered by most to be in the realm of science fiction.

15.2 Early proposals for communication satellites

The first proposals for a communication satellite system were made in 1945 by A.C. Clarke (UK).

A.C. Clarke

Clarke must have been early aware of the space system work being carried out at the UK Royal Aircraft Establishment at Farnborough, England and thus became familiar with the basics of rocketry and satellite orbits. He was later to become famous as a science fiction writer and adviser to the makers of futuristic films such as *2001*.

Clarke's proposals, made 12 years before the launching of the first satellite, appeared in a modest article in a journal, the *Wireless World*, in October 1945. It is perhaps relevant that England, for the previous year or so, had been at the receiving end of the German V2 rocket-propelled missiles, and this may have inspired the idea that there were more useful things that could be done with rockets! [1]

Clarke proposed 'three satellite stations moving in a circular equatorial orbit at a height of 22,300 miles above the Earth's surface, moving in a West to East direction'. At this height the orbital period is 24 hours and the satellites appear stationary when viewed from a west to east rotating Earth. Such an arrangement would provide virtually complete coverage of the globe, except for limited areas near the Poles.

He also envisaged the use of microwaves, noting their freedom from absorption in the atmosphere and ionosphere, with parabolic reflectors to concentrate the transmitted and received energy into narrow beams. Such an arrangement not only greatly reduces the satellite transmitter power required, it also by virtue of the stationary orbit simplifies the problem of pointing the highly directional receiving aerials at the satellite. He foresaw the use of satellites not only for point-to-point telecommunication services such as telephone and telegraph, he also saw the potential for direct broadcasting to homes.

His 1945 article in *Wireless World* refers to the possibility of manned space stations and un-manned satellites 'left to broadcast back to Earth'. He also discusses the use of energy radiated by the sun to power space stations, including the use of thermo-electric and photo-electric devices.

October 1945   **Wireless World**

EXTRA-TERRESTRIAL RELAYS

Can Rocket Stations
Give World-wide Radio Coverage?

By ARTHUR C. CLARKE

ALTHOUGH it is possible, by a suitable choice of frequencies and routes, to provide telephony circuits between any two points or regions of the earth for a large part of the time, long-distance communication is greatly hampered by the peculiarities of the ionosphere, and there are even occasions when it may be impossible. A true broadcast service, giving constant field strength at all times over the whole globe would be invaluable, not to say indispensable, in a world society.

Unsatisfactory though the telephony and telegraph position is, that of television is far worse, since ionospheric transmission cannot be employed at all. The service area of a television station, even on a very good site, is only about a hundred miles across. To cover a small country such as Great Britain would require a network of transmitters, connected by coaxial lines, waveguides or VHF relay links. A recent theoretical study[1] has shown that such a system would require repeaters at intervals of fifty miles or less. A system of this kind could provide television coverage, at a very considerable cost, over the whole of a small country. It would be out of the question to provide a large continent with such a service, and only the main centres of population could be included in the network.

The problem is equally serious when an attempt is made to link television services in different parts of the globe. A relay chain several thousand miles long would cost millions, and transoceanic services would still be impossible. Similar considerations apply to the provision of wide-band frequency modulation and other services, such as high-speed facsimile which are by their nature restricted to the ultra-high-frequencies.

Many may consider the solution proposed in this discussion too far-fetched to be taken very seriously. Such an attitude is unreasonable, as everything envisaged here is a logical extension of developments in the last ten years—in particular the perfection of the long-range rocket of which V2 was the prototype. While this article was being written, it was announced that the Germans were considering a similar project, which they believed possible within fifty to a hundred years.

Orbital Radio Relays

J. R. PIERCE[1]

Bell Telephone Laboratories, Murray Hill, N. J.

While orbital radio relays probably could not compete with microwave radio relay for communication over land, they might be useful in transoceanic communication. Three sorts of repeaters appear to be consistent with microwave art: (a) 100-ft reflecting spheres at an altitude of around 2200 miles; (b) a 100-ft oriented plane mirror in a 24-hr orbit, at an altitude of 22,000 miles; (c) an active repeater in a 24-hr orbit. Cases (a) and (b) require at 10-cm wavelength 250-ft-diam antennas and 100-kw and 50-kw power, respectively; in case (c), using 250-ft antennas on the ground and 10-ft antennas on the repeater, only 100 watts on the ground and 0.03 watt on the repeater would be required; in this case one should probably use smaller antennas on the ground. In cases (b) and (c) the problem of maintaining the correct orientation and position of the repeater is critical; perturbations by the moon and sun might cause the satellite to rock or wander prohibitively.

Figure 15.1   The first satellite proprosals

(A.C. Clarke 1945 and J.R. Pierce 1955)

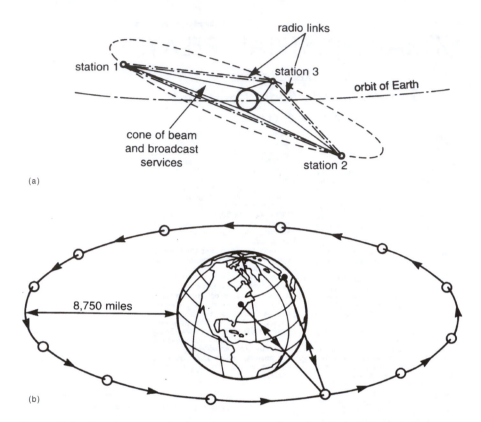

(a)

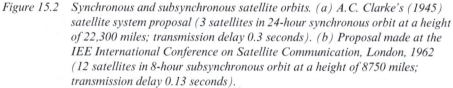

(b)

*Figure 15.2   Synchronous and subsynchronous satellite orbits. (a) A.C. Clarke's (1945)
satellite system proposal (3 satellites in 24-hour synchronous orbit at a height
of 22,300 miles; transmission delay 0.3 seconds). (b) Proposal made at the
IEE International Conference on Satellite Communication, London, 1962
(12 satellites in 8-hour subsynchronous orbit at a height of 8750 miles;
transmission delay 0.13 seconds).*

Clarke's proposals, made so many years before a satellite orbiting the Earth
had been achieved, have all the earmarks of a work of genius for their simplicity,
comprehensiveness and remarkable resemblance to the satellite communication
systems in use today. But, as Clarke has said, 'the publication of the *Wireless
World* article was received with monumental indifference'; he turned his attention and talents to other fields, notably to science fiction writing, in which he
received world acclaim.

Meanwhile, on the other side of the Atlantic Ocean, a scientist at the Bell
Telephone Laboratories – remarkably also a science fiction writer of
considerable repute – had been discussing the possibilities of satellites for
communication, but without knowledge of Clarke's 1945 proposals:

Dr J.R. Pierce (b. 1910)
J.R. Pierce came to Bell Telephone Laboratories after receiving a Ph.D. from California Institute of Technology in 1936, and was Director of Electronic Research at Bell in the 1950s. He was at the forefront of almost every field of research in the telecommunication sciences, from acoustics and information theory to antennas and travelling-wave tubes.

He first put forward his ideas on communication satellites in a talk to the Princeton Section of the American Insitute of Radio Engineers in 1954, later published under the title 'Orbital Radio Relays' in the periodical *Jet Propulsion* in 1955. [2] His paper discusses both passive satellites using spheres, dipoles or plane mirrors as reflectors of radio waves, and active satellites with powered repeaters, in various orbits up to the synchronous orbit at 22,300 miles above the Earth's surface. He also refers to the problems of satellite altitude control and station-keeping that would need to be solved, especially for satellites with directional aerials in the 'stationary' orbit. Pierce's views in 1954 on the utility of satellite communications were remarkably modest: 'while orbital radio-relays probably could not compete with micro-wave radio-relays over land, they might be useful in trans-oceanic communication'.

An article by J.R. Pierce, written under the pseudonym J.J. Coupling and published in *Astounding Science Fiction* in March, 1952, also discusses the possibility of inter-planetary radio communication and terrestrial communication by reflection of radio-waves from the Moon as a passive satellite.

Pierce's realistic approach to satellite communication – which advocated first steps via non-synchronous lower-orbit satellites, typified by the Bell System successful Telstar project of 1961 – was in sharp contrast to the US Army's project ADVENT which aimed in one step at a synchronous-orbit, station-keeping system with all the problems of remote control that this entailed, and which was abandoned in 1962.

With Rudolf Kompfner of Bell Laboratories (see Chapter 11) he drew attention in particular to the vital need for long-life active components, e.g. travelling-wave tubes, for use in satellites if viable commercial operation was to be achieved.

Pierce's book *The Beginnings of Satellite Communication*, published in 1968, [3] gives a lively account of the early history, and contains reproductions of both Clarke's 1945 paper and his own 'Orbital Radio Relays' paper of 1955.

15.3 The Russians are first in orbit, with Sputnik 1

There is no doubt that the launch of the small, grapefruit-sized, satellite Sputnik 1 from Tyuratam in the USSR on 4 October 1957 was a key event that sparked off intensive efforts in the USA to launch satellites, notably but not initially successfully, by the military, who saw important possibilities for surveillance as well as communication. Due to the relatively low orbital height, signals from the low-power VHF transmitter on Sputnik 1 were readily receivable by radio amateurs and others throughout the world; this gave convincing proof of the satellites existence and enabled its orbit to be readily followed.

The successful launch into orbit of Sputnik 1 was a convincing tribute to Russian prowess in rocketry and especially the multi-stage rocket concept attributed to a Russian schoolmaster, Tsiolkowsky.

Konstantin Tsiolkowsky (1857–1953)
By firing the rocket stages in succession as increasing height was reached, and jettisoning each of the initial stages as its fuel was expended, greater orbital heights and pay-loads could be achieved for a given overall expenditure of rocket fuel and launch costs.

15.4 A satellite for all to see: the ECHO balloon satellite

The next significant step was the launching, into a 1,000 mile-high orbit, of a 100ft diameter light-weight balloon satellite, ECHO 1, by the US National Aeronautics and Space Agency. NASA's original intention had been to use the balloon to measure the density of the upper atmosphere by observing the decay of the orbit – information of great importance in the design of future satellite systems.

J.R. Pierce and R.W. Kompfner of Bell Laboratories realised that the balloon satellite with its metallised surface presented a valuable opportunity to explore the reflection of microwaves by a spherical passive reflector and to assess its value for communication puposes. To this end a joint project was established between NASA, the Jet Propulsion Laboratory JPL (then a part of NASA) and the Bell Laboratories. [4]

ECHO 1 was successfully launched into a circular Earth orbit on 12 August 1960 by a Delta rocket, the 1,000 mile-high orbit giving an orbital period of about two hours. Microwave transmissions on 960 MHz and 2,390 MHz were made between the JPL Laboratories at Goldstone, California and Bell Laboratories at Holmdel, New Jersey, using wide-deviation FM to transmit speech signals.

Since the attenuation over the path from the transmitting ground station via the scattered reflection from the satellite to the ground receiving station is very large, low-noise receiving equipment was essential. In the ECHO experiments this was achieved by using a 'maser' liquid-helium-cooled pre-amplifier ('maser' is an acronym for 'microwave amplification by stimulated emission of radiation').

When operated from a steerable horn-reflector aerial of the type originated by Friis and Bruce (see Chapter 11), the maser receiving system was sufficiently low-noise, about 19K, to detect the 3K background noise radiation associated with the 'big bang' and the creation of the universe – an observation for which A.A. Penzias and R.W. Wilson were awarded a Nobel prize in 1978.

Also of interest in the experiments was the use of a 'frequency following' FM receiver, invented by J.G. Chaffee of Bell Laboratories more than 20 years earlier, which lowered the noise threshold of reception by several decibels.

In December 1960, on the first pass of mutual visibility between Goldstone and Holmdel, a voice message from President Eisenhower was successfully transmitted. However, it was clear that passive reflection satellites, with their limited communication capacity – not much more than a single voice channel – had little commercial future.

The real value of the ECHO experiment was two-fold.

First, it was a valuable pre-cursor to the development of the ground station equipment, rocket launching and satellite tracking techniques needed for a commercially viable communication satellite system using active satellites.

Second, it showed the public at large that satellites were 'for real' – no one who watched will forget the sight of the ECHO balloon, glowing in the light of sunrise and sunset, and moving majestically across the sky from horizon to horizon.

15.5 The Telstar and RELAY satellite projects

The Telstar project was created and funded by the ATT Co. and the Bell Telephone Laboratories as a private venture aimed at maintaining their pre-eminence in long-distance telecommunications. The RELAY project on the other hand was initiated by the US National Aeronautics and Space Agency (NASA), with the Radio Corporation of America as the prime contractor. NASA provided rocket launching facilities at Cape Canaveral for both projects and satellite tracking facilities at its Goddard Space Flight Centre. [4]

The primary aims of both projects were the same, i.e. to demonstrate that active communication satellites could provide reliable, good quality long-distance telephone, telegraph, data, facsimile and television circuits – with the prospect of competing in quantity of circuits and cost with established long-distance transmission systems such as land microwave radio-relay and sub-marine coaxial cable systems. A further aim was to collect more scientific data on the space environment as an aid to the design of future satellite systems.

15.6 The Telstar project

The Telstar project was one of the major scientific and engineering achievements of the 20th century, establishing as it did the practicability of world-wide communication by satellites. Whilst the overall planning, design and management of the project were the responsibility of ATT Co./Bell Laboratories, the British Post Office and the French PTT Administration agreed to provide ground stations in their respective countries and to carry out tests and demonstrations in co-operation with BTL.

(a) ATT/BTL 'TELSTAR' Satellite
(34.5 ins diameter, 170 lbs weight)

(b) ATT/BTL Satellite Ground Station at Andover, Maine. USA
(Showing 170 ft long steerable horn aerial and 200 ft diameter
inflated radome)

Figure 15.3   ATT Co./Bell Telephone Laboratories
Telstar Satellite and Andover Satellite Ground Station

The BTL project managers directly involved were A.C. Dickieson, Head of Transmission Systems Development and E.F. O'Neill, Telstar Project Engineer.

Telstar was an active satellite launched in an elliptical orbit varying from 590 to 3,500 miles in height, inclined at an angle of about 45 degrees to the Equator. The inclined orbit enabled a wide sampling of the space environment to be made, e.g. to assess the liability of satellite electronic components such as transistors to damage by the Van Allen radiation belts that surround the Earth, and the possibility of damage from micro-meteorites. The orbital period was about 2.5 hours, the time of mutual visibility for ground stations on opposite sides of the Atlantic being some 30 minutes or less. The ground station aerials were thus required to track the moving satellite as it appeared above the horizon until mutual visibility ceased.

The near-spherical, 34.5 inch diameter, 170-pound Telstar satellite was spin-stabilised, its rotation around a central axis giving it a gyroscopic stability so that its axis tended to remain in a fixed direction in space – thus minimising the risk of loss of signal due to nulls in the directional patterns of the satellite aerials. Radio telemetry was provided for satellite control purposes and to enable the transmission of data from the satellite. The electrical power to operate the communications, telemetry and control equipment was provided by solar cells that converted sunlight into electrical energy stored in nickel cadmium cells both for normal operation and to maintain continuity when the satellite was in the Earth's shadow.

The communication repeater on the satellite provided a bandwidth of 50 MHz, accommodating a frequency-modulated carrier bearing some 600 FDM telephone channels, or a television signal, used a 2 watt output travelling-wave tube and in all the satellite contained about 2,500 solid-state devices, mainly transistors and diodes. The design is a tribute to a Bell Laboratories engineer Leroy C. Tillotson whose internal BTL Memorandum of August 1959 proposed most of the characteristics eventually adopted for the Telstar satellite. [3]

The radio frequencies to be used were agreed in principle by BTL with the cooperating ground station authorities (UK British Post Office and French PTT) on the basis of shared use with the frequency bands already recognised internationally for terrestrial microwave radio-relay systems, i.e. around 4 and 6 GHz. This understanding undoubtedly facilitated the agreement reached later by CCIR Study Group V (Space Systems) and adopted by the Administrative Radio Conference on frequency allocations for operational satellite communication systems, and the conditions to be observed to avoid mutual interference in the shared bands. (In the event, frequencies used for Telstar were 4,170 MHz for the down path and 6,390 MHz for the up path, the lower frequency being used for reception at the ground station because of the slightly lower level of radio noise from the atmosphere.)

The ground station aerial designs differed markedly as between BTL, which used a large steerable horn-reflector under a 200 ft diameter inflated radome at its ground station at Andover, Maine, and the British Post Office with an 85ft diameter open parabolic dish reflector at its Goonhilly, Cornwall ground

Figure 15.4   Post Office satellite communication ground station aerial No.1 at Goonhilly, Cornwall (1962)

station. The French PTT at Pleumer Bodou, Brittany used a replica of the BTL, Andover design.

15.7 British Post Office contributions to the early history of satellite communications

The BPO Engineering Department, notably in its Radio Experimental Branch and Research Department at Dollis Hill, London, had considerable experience in the design and use of microwave radio-relay systems (see Chapter 11) that provided an excellent base from which to initiate work on satellite systems. This had the firm support of the then Engineer-in-Chief, Sir Albert Mumford, who established a team for this purpose led by:

- Capt. C.F. Booth, Asst. E. in C;
- W.J. Bray, Head HQ Space Communication Systems Branch[1];
- F.J.D. Taylor, Head Radio Experimental Branch.

Much consideration was being given in the 1960s to the desirable orbits for a world-wide communication satellite system, taking into account the sometimes conflicting technical, operational and economic factors involved. Early proposals included an initial system of 50 to 100 satellites in random orbit at about 1,000 miles height, put forward by J.R. Pierce of Bell Laboratories, influenced no doubt by the difficulties he then foresaw in achieving reliable station-keeping and altitude control in a synchronous, i.e. stationary, orbit system at a height of 22,300 miles.

Another problem anticipated with the synchronous orbit was the transmission delay of 0.3 seconds which accentuated the effect on talkers of echoes produced at the impedance mis-match frequently encountered at the junction between 2-wire and 4-wire local telephone distribution networks.

This made it difficult for a would-be talker to break-in on one already talking and led the writer to propose a 'sub-synchronous' equatorial orbit with 12 satellites at 8,750 miles height, when the orbital period would be exactly 8 hours, and the transmission delay reduced to 0.13 seconds. Such a system would of course entail tracking the moving satellites, but in a regular and simple pattern. [6]

Eventually the clear advantages of the synchronous orbit and 'stationary' satellites prevailed over the 'break-in' disadvantage, once the problems of accurate station-keeping and altitude control had been solved.

Realising the importance and potential of satellite communications the UK Government Cabinet Office arranged for a joint civil/military team to visit the USA in 1960 to investigate at first hand satellite system developments. The outcome was an agreement with the US National Aeronautics and Space Agency, signed in February 1961, to conduct satellite communication tests across the North Atlantic and an understanding that the British Post Office would build, at its own expense, a ground station for the purpose.

At about this time the BPO was engaged in technical studies with UK military establishments (the Royal Aircraft Establishment, the Army Signals Research and Development Establishment and the Navy Surface Weapons Establishment) to examine the possibilities of a joint civil/military satellite communication system and to consider the possible use of the British 'Blue Streak' rocket with German and French second and third stages, under the aegis of the European Launcher Development Organization, as a means for launching satellites in such a system. Although the outcome was nugatory in that the objectives of civil and military systems proved to be significantly different, and the Blue-Streak rocket project had to be abandoned for a variety of economic, political and technical reasons, the studies themselves proved very educative for the BPO engineers in the fields of rockets, orbits and payloads!. [7]

---

[1] Chapter 14 of my book *Memoirs of a Telecommunications Engineer* contains an account of this exciting period from the viewpoint of one of the participants [5].

15.8   The building of the first UK satellite communication ground station at Goonhilly, Cornwall (1961–1962)

The building of the first satellite communication ground station in the UK in one year, from a virgin site in July 1961 to readiness for the Telstar tests in July 1962, was perhaps an unexpected achievement for a Government Civil Service department such as the British Post Office. However, the BPO Engineering Department had a long history of handling large-scale radio and civil engineering projects, instanced for example by the high-power Rugby Radio Station, which gave it the experience and confidence to 'do its own thing'.

The building of the Goonhilly ground station involved much new and untried microwave technology such as low-noise maser and high-power travelling-wave tube amplifiers, and the construction of an 85ft diameter parabolic reflector aerial weighing over 870 tons, capable of being steered with great precision to track a rapidly moving satellite. It also involved design decisions and the co-ordination of many contributions from various sectors of industry and other Government departments, against a time scale that left no room for errors of judgement. The story of how this was achieved, the design of the station and the results achieved are described in references. [5] [8] [9]

The choice in 1960 of the ground station site at Goonhilly on the Lizard Peninsula, Cornwall was based on the desire to facilitate trans-Atlantic communication – which indicated a location as far west as was practicable – and the need to avoid interference to and from terrestrial microwave radio-relay systems using the shared 4 and 6 GHz frequency bands. Almost by accident the chosen site was near Poldhu from which Marconi made the first long-wave radio transmissions to Newfoundland in 1901 (see Chapter 7).

The civil engineering design and construction of the 85ft diameter parabolic reflector aerial at Goonhilly was due to Dr Tom Husband, whose firm Husband and Co. had been responsible for the 240ft diameter radio-telescope used for radio astronomy at Jodrell Bank, Cheshire. The construction had to be massive to withstand the gales experienced at times on the wind-swept Lizard Peninsula, and yet had to be capable of precision steering – achieved by driving it in azimuth by a substantial Reynolds chain – derided by one newspaper at the time as a 'bicycle chain'! The choice of an open-dish design of aerial, unprotected by a radome as was used by Bell Laboratories at Andover and Plumeur Bodou, proved to be advantageous both in terms of cost and operationally, as the later tests showed.

15.9   The Atlantic is bridged by satellite, July 1962

The Telstar satellite was launched by NASA from Cape Canaveral at 0835 GMT on 10 July, 1962. There followed between the 10 and 27 of July an intensive period of tests and demonstrations that included the transmission of

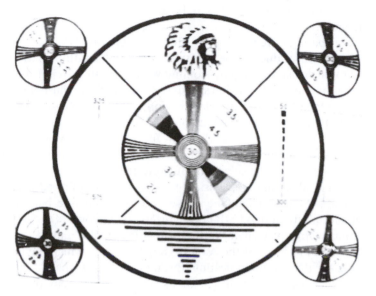

(a) 'Indian Head' Test Card recived at Goonhilly from ATT/BTL Satellite
Ground Station, Andover, Maine from Telstar on 11th July, 1962

(b) Map of UK transmitted from Goonhilly via Telstar on 11th July, 1962

Figure 15.5   The first tests via the Telstar satellite in July 1962

monochrome and colour television signals, multi-channel telephony, data and facsimile signals.

On the first pass of Telstar on 10 July the Goonhilly ground station experienced a difficulty due to the reversal of the polarisation of the aerial, which did not match that of the downcoming wave from the satellite and which arose from an ambiguity in the definition of 'right-handed circular polarisation'.

Thus the honour of the first direct reception of television signals from the USA fell to the French PTT at Plumeur Bodou; however the polarisation error at Goonhilly was soon corrected and on 11 July the first television transmissions to the USA were made from Goonhilly; they were reported by Andover as of 'excellent quality' and re-broadcast throughout the USA.

These exciting and dramatic television exchanges were followed on 12 July by the first Trans-Atlantic telephone conversations by satellite between BPO engineers at Goonhilly and ATT/BTL engineers at Andover. On 14 July Sir Ronald German, Director General of the British Post Office, talked via Telstar with Mr McNeely, President of the ATT Co.mpany during which the President spoke of 'the thrill of communicating through a dot in space' and H.M. Queen Elizabeth herself spoke of Telstar as 'the invisible focus of a million eyes'.

The early transmissions of television across the Atlantic had been in monochrome and some doubt had been felt about the possible effect of Doppler frequency shift on the chrominance sub-carrier of a colour television signal. However, with the help of colour picture generating equipment supplied by the BBC, the first colour television transmissions were made by BPO engineers from Goonhilly on 16/17 July 1962, using 525-line NTSC standards.

The quality of reception at Andover was rated 'very good' and gave further confidence in the transmission quality of the satellite link.

The writer summed up the results of the tests as follows: [9]

The results obtained from the Telstar demonstrations and tests to date (end 1962) have confirmed the expectation that active communication satellites could provide high-quality stable circuits both for television and multi-channel telephony. The very good results obtained with colour television signals – a stringent test of any transmission system – and the tests with 600 simulated telephone circuits, are particularly noteworthy.

15.10 The synchronous orbit is achieved: SYNCOM

To the engineers of the Hughes Aircraft Company in the USA must go the credit for the achievement in July 1963, for the first time, of a satellite in the synchronous orbit at a height of 22,300 miles – thus fulfilling the first stage of A.C. Clarke's vision of 1945.

The Hughes proposals were sponsored by NASA as Project SYNCOM; they involved a satellite of the minimum practicable weight – only 78 pounds – because of the restrictions imposed by the Thor-Delta rocket launcher on the payload that could be put into the synchronous orbit. The need to keep the weight low meant that storage batteries to maintain operation in the Earth's shadow, when the solar cells failed to receive sunlight, could not be included. The SYNCOM satellite contained an apogee (jet) motor for injecting it into the final synchronous orbit after separation of the satellite from the rocket launcher. It was spin-stabilised, and used gas jets on the satellite controlled by radio telemetry from the ground to adjust its position and attitude in orbit.

As viewed from the ground even satellites in the synchronous orbit are not completely 'stationary'; they generally describe a narrow 'figure of eight' pattern above and below a mean position near the Equator, and when high-gain, narrow-beam ground station aerials are used they have to be tracked, albeit over a limited angular range.

The radio equipment on SYNCOM comprised a single repeater, receiving signals from the ground on 7,400 MHz and transmitting on 1,800 MHz with a power of 4 watts from a travelling-wave tube amplifier. Signal-to-noise ratio considerations meant that the repeater bandwidth had to be restricted to about 0.5 MHz; nevertheless, multi-channel telephony, facsimile and data transmission were successfully demonstrated.

The success of Project SYNCOM paved the way for the next generation of satellites by demonstrating beyond doubt that the synchronous orbit was achievable, and that satellite station-keeping and altitude control were practicable.

15.11 The onward march of satellite communications

From the successes of Telstar and SYNCOM the satellite story moves out of the hands of its scientific and engineering innovators and creators – powerful political and financial forces were at work, both nationally and internationally, and the story becomes increasingly complex.

In August 1962 the US Government, realising the immense significance of these developments in satellite communications, passed the Communications Satellite Act, which created a Communications Satellite Corporation (COMSAT) with the aim of securing American control of the manufacture, launching and marketing of satellites for international telecommunications.

It became clear that earlier procedures for the establishment of international telecommunication links such as submarine cables, which rested on bi-lateral agreements between the participating PTT authorities in the countries concerned with ratification by governments, now had to be replaced by multi-lateral agreements which necessarily involved COMSAT. This was recognised by the 22-nation European Conference of Postal and Telecommunication Administrations (CEPT) in December 1962, which set up in 1964 an Interim Communications Satellite Committee in which the USA had through COMSAT a majority ownership share.

The first successful commercial communication satellite launched in April

1965 by COMSAT was the synchronous orbit INTELSAT 11 (Early Bird), which provided 240 telephone circuits or one television channel between Europe and the USA.

From this modest beginning communication satellites were, within two decades, providing the bulk of the world's inter-continental and large numbers of trans-continental telephone and television links. Satellite capacities came to be measured in tens of thousands of telephone circuits and tens of television channels. Specialised satellites for maritime ship-to-ship and ship-to-shore communication, and communication to and from aircraft were launched. And television broadcasting directly from high-power satellites to viewers' homes came into being.

And, far beyond even A.C. Clarke's 1945 prophetic forecast, space craft in the 1980s transmitted, using image storage and slow-scan techniques, high-quality television pictures of the Moon, Saturn's rings and the moons of the furthermost planets of the solar system.

It must have been immensely gratifying for the innovative scientists and engineers involved in the early stages of this remarkable development of the telecommunicators art to have seen their ideas and work come to such fruition within their lifetime.

References

1 A.C. Clarke, 'Extra-Terrestrial Relays: Can Rocket Stations give World-wide Coverage?' (*Wireless World*, October 1945).
2 J.R. Pierce, 'Orbital Radio Relays' (*Jet Propulsion*, April 1955).
3 J.R. Pierce, *The Beginnings of Satellite Communication* (San Francisco Press, Inc., History of Technology Monographs, 1968).
4 *A History of Engineering and Science in the Bell System : Transmission Technology (1925–1975)*, Chapter 13 'Satellites' (ATT Co./Bell Laboratories, 1985).
5 John Bray, *Memoirs of a Telecommunication Engineer*, Chapter 14, 'The Beginnings of Satellite Communication' (Pub. 1982, available via e-mail; bray@btInternet.com).
6 W.J. Bray, 'Satellite Communication Systems' (*Post Office Electrical Engineers Journal*, 55, July 1962).
7 *A Study of Satellite Systems for Civil Communication'* (Royal Aircraft Establishment Report No GW26, March 1961). Published in shortened form in the book *Telecommunication Satellites*, edited by K.W. Gatland (Iliffe Books Ltd, 1964).
8 F.J.D. Taylor, 'Equipment and Testing Facilities at the Experimental Satellite Ground Station, Goonhilly Downs, Cornwall' (*Post Office Electrical Engineers Journal*, 55, July 1962).
9 W.J. Bray and F.J.D. Taylor, 'Preliminary Results of the Project Telstar Communication Satellite Demonstrations and Tests' (*Post Office Electrical, Engineers Journal*, 55, October 1962).

Chapter 16

Pioneers of long-distance waveguide systems: an unfulfilled vision

16.1 Guided electromagnetic waves: the telecommunication highways of the future

The possibility that electromagnetic (EM) waves, whether as microwaves or light waves, might be propagated over long distances on dielectric rods or in metal tubes with or without a dielectric filling was studied from motives of scientific and mathematical interest, long before applications for long-distance comunication seemed either practicable or useful. As long ago as 1897 Lord Rayleigh in the UK had shown, from certain solutions of Maxwell's field equations (see Chapter 2), that electromagnetic waves could propagate freely inside hollow metal tubes provided that the wavelength was appreciably shorter than a cross sectional dimension of the rod or tube. [1] This meant in effect that there was a 'cut-off' frequency below which EM waves would not propagate; for rods or tubes of a convenient size, e.g. a few centimetres across, frequencies of many GHz were required for free propagation. In the absence of suitable sources and detectors for such frequencies, it was not until the 1930s that pioneering experimental studies of centimetric wavelength guided waves began to be made in America by George C. Southworth of the ATT Co./Bell Laboratories. [2] (EM waves on wire pairs or coaxials are also 'guided', but without a lower limit of frequency.) Clerk Maxwell had demonstrated mathematically in the 1870s that light consists of electromagnetic waves, differing only from microwaves in that the wave lengths were extremely short, 0.4 to 0.75 micron (a micron is one ten-thousandth of a centimetre) in the case of visible light, and the frequencies correspondingly high. The term 'light' is conventionally extended to include EM waves in the infra-red region of the spectrum, from 0.75 micron wavelength to the millimetric wavelength microwaves, and also the ultra-violet region below 0.4 micron. Having established the identity of light with EM waves it became clear that in principle light waves too could be guided in glass fibres of diameter greater than a wavelength. And thus there was at least a possibility that 'optical fibres' transmitting visible or infra-red light could be used for communications. However, this possibility was not investigated experimentally until the 1960s when C.K. Kao and G.A. Hockham of Standard Telecommunication Laboratories, England published their first results. [3]

The requirements of inter-city links for multi-channel telephony and television transmission in the period up to the 1980s had been met in Europe, America and most other countries by coaxial cable and microwave radio-relay systems. However, the exponential growth of telecommunication traffic foreseen beyond that period made it worthwhile to intensively study guided-wave systems, with their potentially hugh communication capacity on a single tube or glass fibre. The earliest approaches were to centimetric wavelength guided-wave systems using copper tubes, but as the possibility of achieving low-loss transmission of light in glass fibres began to be realised, the pendulum began to swing in favour of optical systems, with all the operational advantages of a mechanically convenient and compact medium they offered.

16.2 The waveguide pioneers

The accolade for far-sighted vision of the possibilities of waveguides for long-distance communication and a determined attack – in the face of considerable practical and managerial difficulties – on the design problems involved, must go to George C. Southworth.

Assistant J. F. Hargreaves and G. C. Southworth who made the first waveguide experiments, beginning early in 1932

Figure 16.1   Waveguide pioneer G.C. Southworth and waveguide field patterns

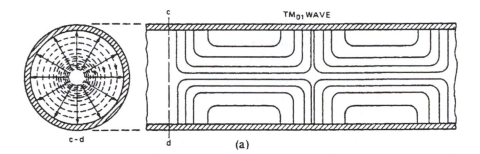

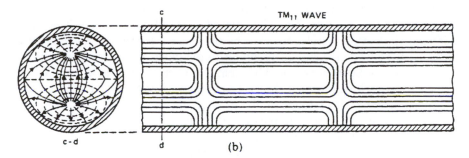

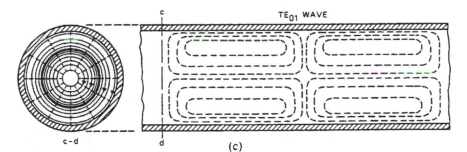

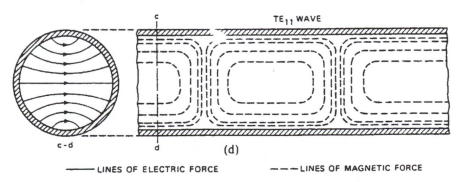

LINES OF ELECTRIC FORCE   — — —LINES OF MAGNETIC FORCE

Electric and magnetic fields in circular waveguide [Southworth, *Principles and Applications of Waveguide /Transmission* (1950): 120.]

Figure 16.1   continued

George C. Southworth

The story of the early waveguide work is well told in Southworth's book *Forty Years of Radio Research*, [2] in which Lloyd Espenschied of BTL has written in the Foreword:

Related are not only the subtle technical problems and how they were met in long-sustained effort, but also the human side of the picture, the motivations, trials, and tribulations in a great stratified organization, periodic victories, and frustrations.

Southworth's first contact with guided-wave phenomena came when, as a student at Yale University in the 1920s, he was experimenting with high-frequency waves propagated in a water-filled rectangular trough and observed unexpected wave patterns. The high dielectric constant of water (c.80) had increased the cut-off wavelength to a point where a variety of wave patterns could exist. These he recognised as similar to the types of wave predicted mathematically by O. Schriever at the University of Kiel in 1920, in a study of the propagation of EM waves on dielectric rods. [4] (As noted above, Lord Rayleigh had much earlier, in 1897, made similar predictions.)

Following a strong instinct to pursue the matter further, and with little initial managerial support, Southworth, then a research engineer with ATT Co., began work on dielectric-filled metal pipes as waveguides in 1931. His notebook of 10 November 1931 contains an entry proposing their use as a transmission medium. The following summary of his subsequent work is based on the account in the Bell *History of Science and Engineering*. [2] [5] [7]

By March 1932 Southworth had set up at Netcong, New Jersey water-filled waveguides six and ten inches in diameter and a few feet long, energised from an oscillator covering 150 to 250 MHz and equipped with launching and detecting devices. With this equipment he was able to identify the various EM modes and measure wave phase and other characteristics.

Remarkably, this work was at first discouraged by the ATT Co. mathematician J.R. Carson on the grounds that 'it was not practical' – perhaps because the laboratories were then struggling with the early coaxial cables and were very much aware that the cable attenuation increased with frequency to a degree which at the time limited the highest usable frequencies to a few MHz.

However, Southworth's experimental results and his determination to take the work further, stimulated a more detailed mathematical analysis of EM wave propagation in hollow metal pipes. This was carried out at ATT Co. by S.A. Schelkunoff and Sally P. Mead.

S.A. Schelkunoff at Bell Laboratories had earlier studied analytically the propagation characteristics of coaxial cables and also turned his attention to guided waves in metal pipes. Mead and Schelkunoff's studies revealed in May 1933 a remarkable property of one particular EM-wave mode – now known as the 'circular-electric TEO1 mode' – in which the attenuation in a truly circular guide decreased inversely as the 3/2 power of the frequency. In practice, surface and other imperfections cause the attenuation eventually to increase with frequency, but only after a very wide useful frequency band has been achieved.

To Sally P. Mead should go the credit for the first formulation of this remarkable result, whilst S.A. Schelkunoff seems to have been the first to realise the significance of the asymptotic behaviour with frequency. [6]

By 1935 a six-inch copper pipe waveguide 1250 ft long had been built at the Bell Laboratory, Holmdel and with it Southworth demonstrated transmission of the TE11 mode wave. However, at that time, the TE01 low-loss mode was above the frequency range of available test equipment and it was not until the late 1930s that the attenuation of this mode could be measured. The measured values were at first considerably above the theoretically predicted values, due to waveguide imperfections that resulted in the conversion of the low-loss TE01 mode to other modes.

It took many years of painstaking research and development before waveguides with losses approaching the theoretically achievable losses with the TE01 mode were achieved. Perhaps the more immediate value of the guided-wave work of the 1930s was the impetus it gave to the microwave radar developments in the war years, and to microwave radio-relay systems in the post-war years.

16.3 Millimetric waveguide development in the Bell system

Work on TE01-mode guided wave transmission was resumed at Bell Laboratories after the end of the war, with the expectation that it would make available a long-distance transmission medium of far wider bandwidth than was offered by coaxial cable or microwave radio systems. It was early realised that, for a practicable system, the waveguide diameter would need to be not more than a few centimetres; it followed that wavelengths in the millimetric range would be required, i.e. frequencies of more than 30 GHz. It was also envisaged that, although other types of modulation were possible, the wide bandwidth available would enable high-bit-rate digital modulation to be used with all the other advantages that this conferred. [8]

During the 1950s the problems of mode conversion from the TE01 mode to unwanted modes, due to waveguide imerfections and bends, were intensively studied. Mode conversion not only increased waveguide losses, it also caused pulse waveform distortion due to the different propagation velocities of the unwanted modes compared with the low-loss TE01 mode. This pointed clearly to the advantages of PCM digital modulation, with its inherent resistance to such distortion.

Improvements in waveguide design were pursued, not only to reduce losses but also with the aim of relaxing the laying requirements with regard to straightness and curvature at bends. Bends were particularly troublesome in untreated copper pipe, due to conversion from the transverse-electric TE01 mode to the transverse-magnetic TM11 mode which had the same wavelength in the guide. One approach was to replace the copper tube with a continuous fine-wire helix which had little effect on the TE01 mode with its almost zero transverse electric field at the circumference, but which absorbed other modes with a relatively strong circumferential electric field. Another approach was to line the copper tube with a dielectric which had a similar effect – and which was cheaper to manufacture. Both these solutions had become available in the early 1960s, but by then the problems of repeater design loomed large.

Early experimental repeater designs used thermionic valves as 'backward-wave' oscillators and millimetric-wavelength travelling-wave tubes as amplifiers, but the close manufacturing tolerances involved indicated strongly the need for more readily manufactured and reliable solid-state devices in a practicable system.

The break through came in 1963 when J.B. Gunn at IBM demonstrated microwave oscillations in gallium arsenide and indium phosphide diodes. Semi-conductor devices developed by Bell Laboratories included the 'IMPATT' (Impact Avalanche Transit Time Diode) as a solid-state millimeter-wave power source, and the 'LSA' (Limited-Space Charge Accumulation) diode. These formed the basis of an all solid-state regenerative repeater, i.e. One capable of accepting weak binary pulses at bit rates of the order of 300 Mbit/sec, re-shaping and amplifying them for onward transmission.

By 1969 the development of waveguide, broadband channel multiplexing components and repeaters had reached a point where a field trial, with the objective of proving the design of a millimetric waveguide system suitable for the large volume of inter-city telecommunications traffic predicted for the 1980s and beyond, could be put in hand. The performance achieved by 1975 was as follows:

- Voice channels per waveguide   238,000*
- Broadband channel bit rate   274 Mbit/sec
- Repeater spacing   31–37 miles
- Waveguide diameter   6 cm.
    * In 124 broadband channels, 62 in each direction

The field trial over a 14 km route used dielectric- lined waveguide with helical mode suppressors at intervals of about 0.5 mile. The overall loss of the wave-guide was remarkably low over a frequency band from 38 GHz to 104.5 GHz, varying from 1 dB/km at band edges to 0.5 dB/km at the centre. [9]

The massive and sustained effort on waveguide system research and develop-ment by ATT Co. and Bell Laboratories had demonstrated beyond doubt that a millimetric waveguide system was practicable, could provide communication capacity far beyond existing coaxial cable and microwave radio-relay systems, and at lower unit cost. Telecommunication authorities in Great Britain, France, Germany and Japan had followed the Bell Laboratories work with great interest and some had put in hand similar development work; the British Post Office Research Department waveguide system field trial is described below.

However, by the 1970s traffic growth in the USA had slackened and a compet-ing system – based on 'optical fibres' – was beginning to make its appearance. Perhaps the last word from the USA on waveguide systems should be left to the pioneer Southworth who in 1962 had written: [9]

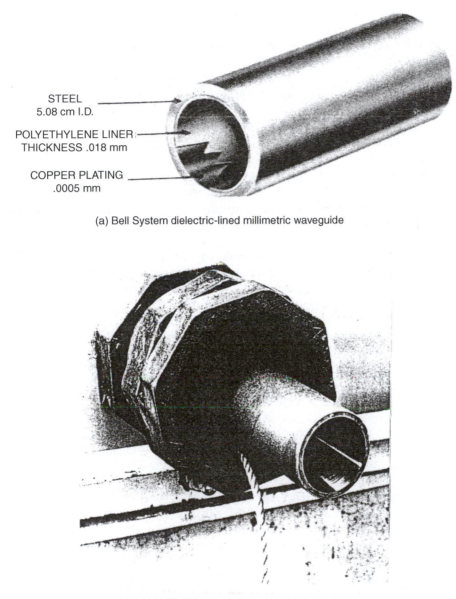

STEEL
5.08 cm I.D.

POLYETHYLENE LINER
THICKNESS .018 mm

COPPER PLATING
.0005 mm

(a) Bell System dielectric-lined millimetric waveguide

(b) British PO/Telecom helix waveguide in duct

Figure 16.2   Waveguide designs for field trials

Almost from the first, however, the possibility of obtaining low attenuations from circular electric waves, carrying with it the possibility of vastly wider bands of frequencies, appeared as a fabulous El Dorado always beckoning us onward.

16.4 Millimetric waveguide development by the British Post Office

Beginning in the middle 1960s scientists and engineers of the British Post Office, initially at the PO Research Department, Dollis Hill, London and later at the PO/British Telecom Research Centre, Martlesham, Suffolk, carried out a comprehensive study of millimetric waveguides operating in the TEO1 mode, leading to a field trial in 1974. In this the BPO was motivated by the same considerations as the Bell System, i.e. to cater in the most efficient way possible for the large and exponentially increasing volume of telephone, data and visual telecommunication traffic between major cities foreseen for the 1980s and beyond. The work received considerable support from UK industry, notably from Standard Telecommunication Laboratories, Harlow in the development of lasers and other solid-state high-bit rate devices, and from British Insulated Callenders Cables who developed manufacturing facilities for the P0-designed light-weight waveguide. A group at University College, London, under the leadership of Professor E.M. Barlow, also contributed with theoretical and experimental studies of guided-wave propagation.

The BPO team on waveguide system research and development was led by Robert W. White and Bill Ritchie.

R.W. White had an extensive earlier involvement in the development of radio-relay and satellite communication systems for the Post Office (see Chapter 15).

Bill Ritchie, who later received a Martlesham Medal for his work on broad-band local distibution systems (see Chapter 19), was responsible for the creation of the unique PO light-weight waveguide. For its field trial the PO opted for an all-helix type of guide in contrast to the Bell Laboratories mainly dielectric-lined waveguide with its rigid steel tube. The 5cm diameter fine-wire helix was supported by an epoxy-resin impregnated glass-fibre tube, giving a precisely-dimensioned, light-weight structure with great mechanical and electrical stability. A critical design factor was the spacing of an aluminium foil boundary around the wire helix, which much reduced mode-conversion losses and yielded a smoother overall loss/frequency characteristic. Also important was the low-loss bend design technique developed by Collin South based on critical dimensioning of the dielectric lining and which enabled losses in bends of only 1 metre radius to be reduced to less than 0.1 db.

A mechanically simple and electrically efficient jointing technique was devised, which enabled sections of waveguide a few metres long to be readily transported and quickly connected under field conditions.

The 5cm diameter waveguide had an effective bandwidth extending from 30 to 110 GHz. Fully equipped with broadband channels operating at 100 to 500

Mbit/sec it provided for up to 300,000 telephone circuits or 200 television channels, or a combination of these.

The field trial, which began in 1974, was over a 14 km route alongside a main trunk road in Suffolk with a variety of straight and curved sections. The light-weight guide was laid in a standard PO cable duct without special supports, its mechanical characteristics enabling it to flex smoothly and gradually. The over-all attenuation of the 14 km length as laid was less than 3dB/km over the 30 to 110 GHz band.

The trial clearly demonstrated that a low-loss millimetric waveguide system was practicable, and that the performance objectives were achievable and stable. It showed that the waveguide could be manufactured in quantity, readily trans-ported and installed under field conditions without excessive constraints as to curvature and bends.

The decision to close down the field trial and further work on millimetric waveguides in 1976 was, in the light of the progress that had been made on optical fibres, a correct decision. Nevertheless, it was a sad moment for the PO/BT scientists and engineers who had laboured so long and successfully to create a practicable and economic waveguide system – at that moment they must have felt a strong bond of sympathy with their friends in Bell Labs who had to accept a similar decision!

But perhaps one useful outcome was the impetus the work gave to the devel-opment of high-bit rate digital techniques and solid-state devices that later found application in optical fibre and millimetric radio-relay systems.

References

1 Lord Rayleigh, 'On the Passage of Electric Waves Through Tubes, or the Vibrations of Dielectric Cylinders' (*Phil. Mag.* Series 5, 43, February 1897).
2 G.C. Southworth, *Forty Years of Radio Research* (Gordon and Breach, 1962).
3 K.C. Kao and G.A. Hockham, 'Dielectric Fibre Surface Waveguides for Optical Frequencies' (*Proc. IEE*, 113, No. 7, July 1966).
4 O. Schriever, 'Elektromagnetische Wellen an Dielektrischen Drahten' (*Ann. der Phys.*, Series 4, 63, December 1920).
5 E.F. O'Neill (ed.), *A History of Science and Engineering in the Bell System, Transmission Technology (1925–1975)*, Chapter 7. 111 Microwaves and Waveguide, (ATT Co./Bell Laboratories, 1985).
6 J.R. Carson, S.P. Mead and S.A. Schelkunoff, *Hyper-Frequency Waveguides – General Considerations and Theoretical Results*, (*Bell Syst. Tech. Jl.*, 15, April 1936)
7 G.C. Southworth, *Hyper-Frequency Waveguides – General Considerations and Experimental Results*, (*Bell Syst. Tech. Jl.*, 15, April 1936).
8 Op. cit. Ref. [5], Chapter 20, III Millimeter Waveguide.
9 'Millimeter Waveguide System', Special Issue *Bell Syst. Tech. Jl.*, 56, Decem-ber 1977).

Chapter 17

Pioneers of optical fibre communication systems: the first trans-Atlantic system

17.1 The objective and the challenge

The objective facing the pioneers was the development of a transmission system offering greater communication capacity at lower cost than existing media such as coaxial cable and microwave radio-relay systems, in a mechanically flexible and compact form that could readily be used in the field.

Hair-thin glass fibres transmitting coherent light offered the possibility of almost unlimited communication capacity. Light with a wavelength of the order of 1 micron has a frequency of 300 million MHz – if only one per cent of this frequency is effectively used the corresponding communication bandwidth is 3 million MHz, wide enough to accommodate millions of telephone circuits or hundreds of television channels on a single fibre. Furthermore, the use of light as a carrier meant complete freedom from electromagnetic interference such as that originating from electrical machinery, power lines and radio transmissions.

The technological challenges involved were substantial – first was the development of low-loss glass, starting from an initial level of 1,000 db/km or more, reducing it to a useable 10 dB/km and eventually to 1 dB/km or less. And this had to be achieved in a fibre structure in which the refractive index distribution between an inner core and an outer glass cladding effectively guided a light wave.

Means had to be developed for continuously drawing glass fibres in lengths of tens of kilometres or more as a basis for a practicable and economic manufacturing process. The next challenge was the development of solid-state laser light sources and photo-detectors, suitable for operation at pulse bit rates of 100 Mbit/sec or more, with the long lives needed for commercial operation.

The possibilities of optical fibres began to be realised in the late 1960s, stimulated by the pioneering work of C.K. Kao at Standard Telecommunication Laboratories, England. By the early 1970s the research laboratories of the British Post Office, Dollis Hill and STL in England, the American giant Corning

Glass, ATT Co. and ITT Co in the USA, and NTT in Japan had begun intensive studies of the problems of creating a viable optical-fibre system technology.

17.2 The first step

The classic paper by C.K. Kao and G.A. Hockham in 1966 presents the first comprehensive scientific, mathematical and experimental study of the transmission of EM waves at optical frequencies by a circular section dielectric fibre with a refractive index higher than the surrounding medium – as such it provided not only a stimulus to further work but a sound foundation for it. [1]

Dr Charles K. Kao (1933)

Dr Kao was born in Shanghai, China in 1933. He gained B.Sc. and Ph.D. degrees at London University and joined Standard Telecommunication Laboratories, England where his work on optical fibres was carried out. His analysis of wave propagation follows that of Lord Rayleigh in 1897 in its use of Maxwell's equations (see Chapter 2) to define a family of magnetic H and electric E modes, and a family of hybrid HE modes.

In particular Kao identified and demonstrated experimentally the lowest order hybrid HE11 mode with which it is possible to achieve single-mode operation by reducing the diameter of the guiding structure. This was later to prove

Figure 17.1   Charles Kao, who with G.A. Hockham led the way to optical fibre systems (1966)

significant in achieving high bit-rate transmission, as compared with multi-mode operation which led to pulse waveform distortion and lower useable bit-rates.

The paper goes on to discuss energy losses due to radiation and bending, scattering from imperfections and absorption from impurities in the glass; it explores the variation of these losses with wavelength and reveals the existence of absorption bands due to molecular resonances in various types of material. In the case of fused quartz for example his results indicated an optimum wavelength of about 0.65 microns. (Later work indicated even lower losses at about 1.3 microns.)

Kao's paper does not give measured losses for optical fibres; however, due to the imperfections of glasses available at the time, these were probably of the order of 1,000 dB/km. He concludes: [1]

1. Theoretical and experimental studies indicate that a fibre of glassy material in a cladded structure with a core diameter of about 1 (the light wavelength) and an overall diameter of 100 represents a possible practical waveguide with important potential as a new form of communication medium.
2. The refractive index of the core needs to be about 1 % above that of the cladding.
3. The realisation of a successful fibre waveguide depends, at present, on the availability of suitable low-loss dielectric material. The crucial material problem appears to be one that is difficult, but not impossible, to solve. Certainly the required loss figure of around 20 dB/km is much higher than the lower limit of loss imposed by fundamental mechanisms.

Time has shown that Kao's concepts of the optical waveguide and his expectations for it were not only sound but have been realised in full measure in operational systems throughout the world.

17.3 The Corning Glass contribution

It was to be expected that American Corning Glass – the largest commercial organisation in the world concerned with the development and manufacture of glass and glass products – would take an early interest in the possibilities of optical fibres.

This interest began in 1966 – significantly the date of publication of Kao's paper – when a Corning scientist William Shaver visited London and discussed the possibilities of optical fibres for telecommunications with engineers of the British Post Office. The story of what flowed from that beginning is well told in an extensive interview with Corning staff in 1989 by a consulting firm concerned with the evolution of strategy to protect American global business interests. [2]

The task facing Corning was daunting indeed – the light loss in even good quality glass was some 20 dB per metre, compared with the objective of not more than 20 dB per kilometre. It was a task that took more than 10 years to complete to the stage of a pilot manufacturing plant capable of making kilometre lengths of glass fibre of acceptable quality – an achievement that involved

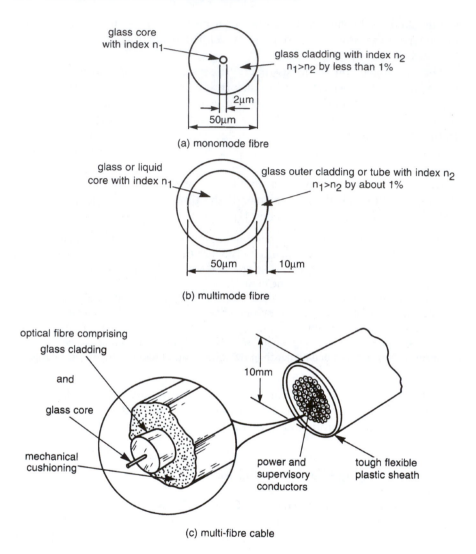

glass core
with index n_1

glass cladding with index n_2
$n_1 > n_2$ by less than 1%

2μm

50μm

(a) monomode fibre

glass or liquid
core with index n_1

glass outer cladding or tube with index n_2
$n_1 > n_2$ by about 1%

50μm   10μm

(b) multimode fibre

optical fibre comprising
glass cladding

and

glass core

mechanical
cushioning

10mm

power and
supervisory
conductors

tough flexible
plastic sheath

(c) multi-fibre cable

Figure 17.2   Single and multi-mode optical fibres

not only a long-sustained and determined effort in the face of many setbacks, but an original approach to fibre design that differed markedly from competing designs. Much of the credit for this must go to Robert Maurer. He was born in Arkansas in 1924 where he gained a BS degree in physics in 1948, followed by a Ph.D. at MIT in 1951. He joined Corning in 1952 as a physicist, later becoming Manager for Applied Physics Research and Special Projects, and Research Fellow. In addition to many technical publications on optical waveguides, he holds some 16 patents including the two key patents in current use throughout the world. His honours and awards include the Morris Liebmann Award of the

Figure 17.3   Three scientists at Corning Glass USA who developed the first low-loss optical fibre
(Left to right: Dr D.B. Keck, Dr R.D. Maurer and Dr. P. Schultz)

US Institution of Electrical and Electronic Engineers, the L.M. Ericsson International Prize for Telecommunications awarded by the Swedish Academy of Engineering, and the American Physical Society International Prize for New Materials; he was elected to the US Academy of Engineering in 1979.

His principal colleagues in the optical fibre work were: Pete Schultz and Donald Keck.

Schultz was a senior Corning scientist who specialised in creating new chemical formulae for glass, whilst Keck tackled the problems of measuring fibre purity and drawing it into a fibre. With Maurer they made a determined and capable research team.

Following his visit to the British Post Office engineers in 1966 William Shaver put the problem of creating a commercially viable optical fibre to William Armistead, Corning's Director of Research. Armistead realised the massive difficulties involved but saw the long range potential of optical fibres and gave Maurer the go-ahead, realising that there was mounting competition from other laboratories in England, W. Germany and Australia as well as the USA.

Corning had gained earlier experience from defence projects with fused silica, the purest glass then known. But it seemed at first most unpromising for an optical waveguide. First, the melting point was extremely high – some 2,000 deg.C, more than twice the melting point of steel – and this would make drawing into a fibre very difficult. Second, the refractive index of fused silica was the lowest of known glasses. And as Kao had pointed out, the fine inner glass core of an optical waveguide has to have a refractive index slightly greater than that of the glass cladding, if light is to be effectively guided.

Whilst Corning's competitors – including the British Post Office research team – were concentrating on refining the impurities out of optical glass, Maurer decided to stay with fused silica, a decision which later proved to have been critical for ultimate success. [3]

The key to the refractive index problem was chemically to dope the material of the inner core to increase its refractive index. As was remarked at the time, while the competition was refining chemicals out, Maurer was putting them in!

There remained the problem of drawing glass into a hair-thin fibre with the right dimensions of inner core and cladding. The well-tried solution used by most of the competition was to take a glass rod, surround it with cylindrical glass cladding of lower refractive index, heat the two in a vertical furnace and draw from the lower melted tip – a process which maintained the relative dimensions of core and cladding but which was difficult to achieve with fused silica. The brilliant solution put forward by Schultz and Keck was to turn the process inside out – to start with a hollow tube of fused silica and apply the fused silica core material by vapour deposition, a technique which Corning had used successfully in other applications.

In April 1970, after four years of endeavour and many frustrating set-backs, a break-through was achieved and Donald Keck was able to demonstrate to Bill Armistead, Corning's Director of Research, a metre length of laboratory-made fibre with a loss below the then target of 20 dB/km; Keck's laboratory notebook says:

Attenuation equals 16 dB. Eureka!

As Keck said at the time,

he could just about feel the spirit of Edison in the laboratory.

But this, vital as it was, was only a beginning; techniques for manufacture of long lengths of fibre had to be developed and a market found amongst the telecommunication equipment manufacturers of the world.

It fell to Charles Lucy, another MIT graduate and who joined Corning in 1952, to supervise the building of an optical fibre manufacturing plant capable of producing long lengths of fibre in a form suitable for use by telecommunication system designers. He had also to find a market for the product.

There was at first a marked reluctance to buying Corning fibre by overseas telecommunication equipment manufacturers and their cable suppliers, although a number of patent licensing greements were made with firms in England, France, Italy and Germany.

Meanwhile Corning continued their research, achieving 4 db/km in 1975, in fibre lengths up to some 4 km. Strenuous efforts were made to improve the manufacturing process, reduce costs and increase the fibre length. By 1977 a new manufacturing process had been achieved in which a rod was sprayed with vaporised core glass and cladding sprayed on that; the rod was then removed and the fibre pulled from the double cylinder that was left. This process lent itself to the production of single-mode fibre with its ultra-fine core – an essential for achieving the maximum communication capacity.

Furthermore, the loss had been reduced to an incredible 0.3 dB/km. By 1988 the manufacturing plant was capable of producing up to 100 km lengths of single-mode low-loss fibre at a very competitive cost – and the door was open to the world market. However, the achievement of a majority share of that market involved much litigation against overseas and some US firms for infringement, litigation in which Corning was notably successful through its ownership of optical fibre master patents.

The Corning optical fibre story is another illustration of the value of research carried out by dedicated individuals with a clear vision of the future, when backed by management which shared that vision, and the both physical and financial resources that a large organisation could provide.

17.4 Light sources and detectors for optical fibre systems

An optical fibre communication system requires a light source which can be focused into a narrow beam for injection into the fibre and which is capable of high-bit rate modulation, and a receiver, i.e. a photo-detector, responsive to low-energy light. Ideally the light source is 'coherent', that is, it is confined to a single frequency and the phase is non-random; in practice there are many light sources with spectrum line widths of much less than one per cent. The coherence of the light emitted by the source is an important factor in concentrating the light into an intense narrow beam; it also facilitates efficient use of the spectrum , e.g. by high bit-rate modulation and light carrier-frequency multiplexing.

The invention of the 'laser' (Light Amplification by Stimulated Emission of Radiation) by A.L. Schawlow and C.H. Townes of Bell Laboratories in 1958, and the first realisation of laser operation by T.H. Maiman of the Hughes Aircraft Company in 1960, provided an important first step. [4] [5]

The basic operation of a laser consists of pumping atoms of a gas or solid into a stimulated state from which electrons can escape and fall to lower energy ground states by giving up photons (light) of the appropriate energy. By providing such a device with a pair of parallel mirror surfaces several hundreds of odd-number half-wavelengths apart, continuous oscillations can be generated by an avalanche action. The pump energy may be from microwave, light or other electromagnetic wave sources.

The helium-gas laser, operating at a wavelength of about 1 micron, was a useful near-coherent light source for laboratory experiments, but was not practicable for a communication system.

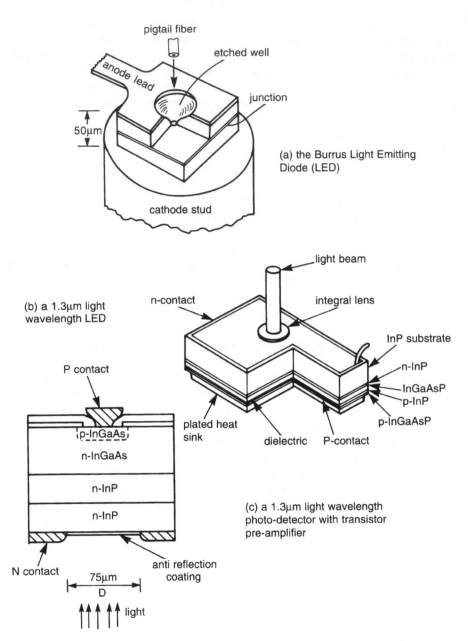

pigtail fiber

etched well

anode lead

junction

50µm

(a) the Burrus Light Emitting
Diode (LED)

cathode stud

light beam

(b) a 1.3µm light
wavelength LED

n-contact

integral lens

InP substrate

n-InP

InGaAsP
p-InP

P contact

p-InGaAsP

p-InGaAs

n-InGaAs

plated heat
sink

dielectric   P-contact

n-InP

n-InP

(c) a 1.3µm light wavelength
photo-detector with transistor
pre-amplifier

N contact

75µm

D

anti reflection
coating

light

Figure 17.4   Light sources and photo-detector for optical fibre systems

The invention of the transistor stimulated the search for a semi-conductor light source in Bell Laboratories and other laboratories during the 1960s. Early versions could only be operated at low temperatures and with intermittent operation. A break through came in 1970 when Bell Laboratories first achieved continuous operation of a semi-conductor laser at room temperature [6]. Although much development lay ahead it was clear that a semi-conductor device – of millimetre dimensions – could provide a near-coherent light source suitable for an optical fibre communication system. [6]

Parallel research on 'light-emitting diodes' (LEDs) produced a light source which, although not highly coherent (the line width at wavelengths of some 0.8 microns being typically about one per cent), nevertheless could be formed into a sharply focused beam from a very small area. Moreover the LEDs could be readily modulated at bit rates of up to 10 Mbit/sec or more; they were thus well adapted to systems of modest capacity and length, operating with multi-mode fibres. One such device, with a single semi-conductor junction, designed by C.A. Burrus of Bell Laboraties in 1971, emits light in a direction normal to the plane of the junction. Another 'stripe edge' geometry results in light emission from the side of the junction. Yet another design – the 'injection laser' diode – is similar to the Burrus diode but incorporates a gallium arsenide layer flanked by n- and p-type layers doped with indium phosphide. With this design it is possible to achieve an operating wavelength of about 1.3 microns which corresponds to a very low-loss region of the light spectrum in optical fibre. [7] [8] [9]

Development of light-sensitive photo-diodes of the avalanche type resulted in a modern Bell Laboratories design termed a 'InGaAs PIN'; it incorporates a window with an anti-reflection layer, backed by an indium-phosphide n-type photo-sensitive layer and an integral field-effect FET GaAs transistor pre-amplifier, designed for an operating wavelength of 1.3 microns. [8]

The physics underlying the creation of the light sources and detectors that are as essential to optical communication systems as the fibre itself, can be traced back to the atomic model delineated by J.J. Thomson and Ernest Rutherford, and extended by Niels Bohr (1885–1962). The further development of the quantum theory by Max Planck, Schrodinger, Fermi, de Broglie, Niels Bohr and others not only revealed much about the behaviour of electrons, waves and photons in the atomic world, it provided an indispensable basis for the design of optical devices (see Chapter 2). [10]

17.5 Early Bell system optical fibre system field trials

During the 1970s the Bell Laboratories continued their development of long-life solid-state light sources and detectors, and optical fibre drawing techniques. The latter achieved in 1974 a significant advance with a chemical-vapour deposition technique, using a hollow tubular glass preform. It will be recalled from earlier in this chapter that Corning were also successfully using a vapour deposition technique inside a tubular glass structure. [9]

A strong motivation for the first field trial of an optical system by Bell was to make better use of existing duct space to meet economically the ever-growing demand for inter-exchange telephone circuits in the local area. For such purposes the optical fibre system with its small size, low losses and large circuit capacity offered an attractive solution.

The first field trial, carried out jointly by Western Electric and Bell Laboratories, was implemented at Atlanta, Georgia in 1975. The cable in this ambitious experiment carried 144 optical fibres with a loss of less than 6dB/km, permitting a spacing of regenerative repeaters of at least 10 km. Each fibre operated at a bit rate of 45 Mbit/sec, using a GaAs laser light source at 0.82 microns wavelength. Thus the complete cable had a potential telephone circuit capacity exceeding 100,000 telephone circuits.

During the period up to 1980 smaller cables with 24 fibres each operating at 45 Mbit/sec were developed and used operationally for telephone and video circuits. By February 1983 an inter-city trunk system operating at 90 Mbit/sec had been installed between Washingtom and New York over a distance of 250 miles.

Thus by the mid-1980s optical fibre systems had proved themselves both in the inter-exchange local area and for inter-city trunks. The success of these inland systems paved the way, and gave confidence for, an attack on the next major objective – a trans-Atlantic optical fibre cable system.

17.6 Optical fibre research and development in the UK

Interest in the possibilities of optical fibres for civil telecommunications, based initially on theoretical studies, began in about 1965 at the Research Department of the British Post Office, Dollis Hill, London. At the same time the UK Government Signals Research and Development Establishment (SRDE) examined possible military applications. However, it was undoubtedly Charles Kao's work at STL and classic paper of 1966 [1], outlined earlier in this chapter, that stimulated more detailed scientific studies and experimental work by the BPO at Dollis Hill. Thus began an investigation in depth of optical fibre systems that included the mechanisms of guided-light propagation in structured glass fibres and the losses therein, and the solid-state light sources and detectors needed for a communication system. BPO studies on optical fibre systems and devices up to the mid-1970s have been described in a series of papers in the *Post Office Electrical Engineers Journal*. [11] [12] [13] [14]

The group at Dollis Hill initially responsible for this investigation was led by the scientists F.F. Roberts, J.I. Carasso, J. Midwinter and Dr M.M. Faktor; their work was continued and expanded at the BPO Research Centre, Martlesham, Suffolk – opened in 1975 (now British Telecom Research Laboratories). A member of this pioneering group, Dr Marc Faktor (1930–1988) was awarded the BT Martlesham Medal in 1987 for outstanding scientific achievement and his pioneering work in opto-electronic techniques in the 1970s and 1980s. Much of his work was concerned with crystal-growing from vapour, that is, building up epitaxial layers of semi-conductor material no more than a millionth of a centimetre thick – a process vital to the creation of light sources and detectors. In particular he pioneered a metal-organic vapour-phase epitaxy (MOVP) technique that was later exploited on a commercial scale jointly by BT and duPont, in a world-wide market for opto-electronic components and devices valued at more than £4 billion. In the early 1970s he headed a team which used organic materials to build long-life semi-conductor lasers which could be directly modulated. He also devised the electro-chemical technique which is the heart of the BT 'profile plotter' for displaying the electric-current carriers in semi-conductor materials – a device which has been manufactured under licence and sold world-wide. [15]

Marc Faktor was born in Poland in 1930 and came to the UK as a schoolboy in 1946, after grim war-time experiences. He began work in a London tutorial college and gained a first degree by evening class study at Birkbeck College, London University, followed by a Ph.D. in inorganic chemistry in 1958 when he joined the BPO Research Department at Dollis Hill. He retired from BT in 1982, and became until his death a Visiting Professor at Queen Mary College, University of London.

He was a modest man of great personal charm who made, by his scientific skill and dedication, an enduring and major impact on the development of optical fibre communication systems.

17.7 Research, development and field trials of optical fibre systems by British Telecom

The move of PO research from Dollis Hill to the new BT Research Centre at Martlesham, Suffolk in 1975 with its improved facilities, and the challenges to a 'privatised' BT from its competitors Mercury and Cable and Wireless, created both the capability and the incentive to pursue the development of optical fibre communication systems with determination and substantial resources. The continuing growth of telephone traffic and the introduction of new telecommunication services required increased capacity on junction and inter-city trunk routes, and this had to be achieved at the lowest practicable cost.

Similar requirements appeared on the submarine cable routes between the UK and European countries, and across the Atlantic, to provide a balance with the growing capacity on satellite systems. And there were potentially important new applications for optical fibre systems in the local distribution network to customers' premises for cable television and enhanced telecomunication services. There were several technological problems to be resolved, ranging from the need to ensure a large-scale and economic source of supply of the optimum design of stable, low-loss fibre cable, to the practical problems of jointing cables in the field and inserting them in existing cable ducts. Designs for vital components such as long-life light sources and optical detectors suitable for operation at high bit-rates had to be established. And overall system engineering studies were needed to ensure the most economic design for each application. Economic and reliability considerations pointed firmly to the advantages of the lowest practicable fibre loss with the widest possible spacing of repeaters, and to mono-mode fibre operation with its large communication capacity where this could be effectively utilised, e.g. on trunk routes. On the other hand multi-mode fibre, with its larger core and lower manufacturing costs, had advantages for smaller capacity systems, e.g. in the local distribution network. And the ultimate goal of a non-repeatered trans-Atlantic optical fibre submarine cable provided a challenging longer-term objective.

The following summary of BT research and development in optical fibre communication systems, part of which was carried out in collaboration with the UK telecommunications industry, outlines the more important technological milestones along the way and points to some of the pioneers whose achievements made these possible.

17.7.1 Low-loss optical fibre research

From earlier work in UK, US and other laboratories two distinct optical fibre glass systems emerged. The one chosen for the UK-based effort, including the PO/BT laboratories, involved low melting-temperature glasses such as the sodium calcium silicates (akin to window glass, but ultra-pure) and the sodium boro-silicates (of which Pyrex is one). These offer a wide range of optical, mechanical and chemical properties but, being melted in crucibles from powders, extreme care has to be taken to avoid impurities.

The other glass system, developed in the USA notably at Corning (see earlier in this Chapter), uses pure fused-silica (the glass form of quartz) and silica doped chemically to produce glass of a slightly different refractive index for the core material. The fused silica system has led to remarkably low losses – approaching 0.2 db/km at 1.3 microns wavelength.

The pulling of molten glass into a fibre with the requisite structure and purity for low-loss light transmission, in unbroken lengths of 10 km or more, is a difficult and demanding task. One successful method uses the 'double crucible' process, evolved and developed into a fibre manufacturing technique by two BT scientists: Dr George Newns and Dr Keith Beales.

Their apparatus uses an inner platinum crucible to contain the molten core glass, and an outer crucible containing molten cladding glass with a lower refractive index. When electrically heated, molten glass begins to flow through concentric nozzles and can be pulled into a filament wound on to a drum as it cools and solidifies. By careful control of temperature distribution and winding speed the correct dimensions of core and cladding can be achieved. By choice of nozzle dimensions either mono-mode or multi-mode fibre can be made; the process is particularly well suited to the low-cost production of multi-mode fibre with its relatively large inner core. It also enables graded index fibres to be produced, the latter being achieved by a diffusion process between core and

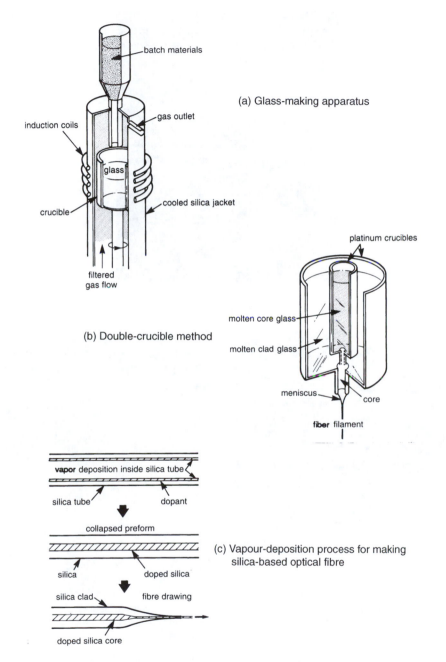

Figure 17.5   Optical fibre glass-making methods

Dr Marc Faktor
MM (1987) for pioneering work in opto-electronics

Dr George Newns   Dr Keith Beales
MM (1982) for innovation in optical fibre technology, including the double-crucible fibre
manufacturing process

*Figure 17.6   Martlesham Medal winners for achievement in optical fibre transmission
systems*

cladding before solidification occurs. The Japanese 'Selfoc' fibre is made in this manner. The two-crucible fibre production process originated by George Newns and Keith Beales has been licensed to British, American and European companies. Newns' and Beales' later work included further development of the vapour-deposition process using silica glass preforms, similar to the Corning process described earlier in this chapter.

For their outstanding contributions to glass-fibre technology Dr George Newns and Dr Keith Beales were awarded the Martlesham Medal in July 1982. [16]

George Newns, who was born in 1936, gained a first-class honours B.Sc. degree from Liverpool University, followed by a Ph.D. He joined the BPO in 1962, and after working on semi-conductors he was appointed in 1968 to lead a research team investigating optical materials, and later fibre fabrication and evaluation for optical communications. He has published over 50 scientific papers and holds a number of patents. With Dr Beales and other colleagues he was awarded the Institution of Electrical Engineers (London) Scientific Premium for a paper on the double-crucible fibre process, and in 1980 the Potts Medal by the Chemical Society of Liverpool University.

Keith Beales, who was born in 1940, won an Open Scholarship to Nottingham University where he gained an honours B.Sc. degree in chemistry and a Ph.D. for a thesis on organic semi-conductors. He joined the Research Department of the BPO in 1966 where his work involved the materials aspects of semi-conductor devices and optical fibres. Since 1979 he has led the research and development on optical fibres, including the double-crucible process, mono-mode fibre development, and fibre strength and reliability studies. He has published over 50 scientific and conference papers and filed 15 patents.

Mention should also be made to the valuable role played by the Optoelectronics Research Centre at the University of Southampton under Professor W.A. Gambling, F.Eng., FRS. The Southampton team developed in the 1970s a process for making low-loss fibre based on phosphorus doped silica, using technology similar to that invented at ATT/Bell by J. MacChesney (Modified Chemical Vapour Deposition). Much good work on optical fibre assessment and propagation studies has since been carried out over the years at Southampton University, including the development of special fibres for sensors.

To British Telecom Research at Martlesham should go the credit for their early recognition of the longer-term advantages of single-mode, as opposed to multi-mode, fibre; their work on the development of single-mode fibre was acknowledged by the Queen's Award to the fibre section of BT Research in 1985.

17.7.2 British Telecom optical system field trials

By the early 1970s the development of optical fibres and devices needed for a communication system had progressed to a point where field trials could be envisaged. The emphasis was initially on the use of multi-mode fibre operating at a wavelength of about 0.8–0.9 microns and which, by virtue of its relatively large core diameter, could be more efficiently coupled to LED light sources and to detectors. It was also more readily readily spliced (joined) under field conditions than mono-mode fibres.

At the opening of the British Telecom Research Centre at Martlesham in 1975 an 8 Mbit/sec 6 km laboratory-based optical fibre link was successfully demonstrated using multi-mode fibre at about 0.8 micron wavelength.

In 1977 field trials using graded-index multi-mode fibre operating at 140 Mbit/sec and installed in ducts in the Martlesham area gave the first demonstration that fibre cables could be installed in ducts, spliced where required, and operated over repeater sections longer than on conventional copper cables. The 140 Mbit/sec rate meant that a single fibre could provide nearly 2,000 telephone channels or a high quality television channel.

By the late 1970s the scene was changing dramatically. Telecommunication research laboratories throughout the world were drawing attention to the lower fibre losses to be achieved by operating at wavelengths of about 1.3 or 1.5 microns, and the superior communication capacity offered by mono-mode fibres. Optical device and component research and development began to be diverted to this new region of the light spectrum.

In 1980 BT Research Laboratories demonstrated a 140 Mbit/sec 1.3 microns wavelength optical fibre link using components developed at the laboratories and operating over an un-repeated 37 km length of mono-mode silica cable. This was a crucial demonstration since it meant that the power-feeding sections in the existing BT trunk network, spaced up to 30 km, could now be bridged by an unrepeated section of optical fibre cable – a result which gave a substantial economic advantage to the optical fibre system compared with conventional coaxial cables. [17]

Also in 1980 a 10 km submarine cable link, using identical components to the land system referred to earlier, was demonstrated at Loch Fyne, Scotland in collaboration with Standard Telecommunication Laboratories.

During 1982 two significant field trials of duct-installed single-mode optical fibre cable operating at 1.3 microns wavelength were carried out by BT in Suffolk, partly with a view to examining the effect of cable tolerances on splicing losses. The first, carried out in co-operation with Telephone Cables Ltd, involved the installation of a 7.5 km route near Woodbridge, Suffolk using experimental fibre made by BTRL and GEC Optical Fibres Ltd. The second, made in association with STL, was over a 15 km route between Martlesham and Ipswich using cable made by STL/STC in a pilot manufacturing plant. The splicing loss tests revealed clearly the benefits of the closer dimensional tolerances of the second cable.

The 'Woodbridge' cable was operated initially at 140 Mbit/sec and later at 650 Mbit/sec; the 'Martlesham–Ipswich' cable was operated at 565 Mbit/sec over a looped distance of 61.3 km, an achievement believed to be a world first.

Additional tests at a wavelength of 1.55 microns achieved a bit rate of 140

Mbit/sec over an un-repeatered looped distance of 90.5km on installed Martlesham–Ipswich cable. [17]

These field trials created a sound basis for proceeding in the UK with a nation-wide junction and trunk network based on mono-mode optical fibre systems; they also gave confidence that there would be an important role for optical fibre submarine cables, even spanning the Atlantic Ocean.

By 1982 British Telecom had several thousand kilometres of first generation 0.8 micron wavelength optical fibre cable systems in service. In July 1982 a second generation, 1.3 microns wavelength 140 Mbit/sec, repeatered optical fibre link 204 km long – then the world's longest optical link – went into service between London and Birmingham. This link used cable made by BICC Tele-communication Cables Ltd and equipment made by Plessey Telecommunications Ltd to BT specifications. [18]

And a new name was coined by British Telecom for its optical fibre cable systems – 'Lightlines'.

By 1986 there were more than 65,000 km of second generation 140 Mbit/sec optical fibre cable systems in use in the BT network. In that year two 565 Mbit/sec optical fibre systems were also put into service – one of which used ultra-reliable opto-electronic and microchip components designed and made by BTRL as a test-bed for component designs to be used in a trans-Atlantic optical fibre cable (TAT 8). [18]

17.7.3 Optical fibre cable systems go under the ocean

Following the successful optical-fibre submarine cable experiment in Loch Fyne in 1982, BT initiated plans for operational links across the North Sea and to Ireland.

The world's first international optical fibre submarine cable link, between the UK and Belgium, was opened in October 1986 with a two-way London–Ostend video conference. The cable system was supplied by STC Submarine Cables Ltd to BT order and specification, the main 80 km deep water section being laid by the British Telecom International cable ship *Alert*. The optical fibre system provides more than 11,500 additional telephone circuits across the North Sea.

In 1988 optical fibre cables were laid across the Irish Sea, one 90 km long between Anglesey and the Isle of Man, and another 126 km long between Anglesey and Portmarnock in the Irish Republic. Neither of these cables required intermediate repeaters on the sea bed – a clear illustration of the remarkable success achieved by the long scientific and engineering battle to reduce losses in optical fibres.

They paved the way for an even more dramatic success in the optical fibre story – the spanning of the Atlantic Ocean by an optical fibre cable system.

A comprehensive collection of papers on optical fibre submarine cable system development has been given in the *Journal of British Telecommunication Engineers*. [19]

17.8 The first trans-Atlantic optical fibre submarine cable system (TAT 8)

The world's first trans-Atlantic optical fibre cable system, TAT 8, capable of carrying 40,000 simultaneous telephone conversations, was opened for service between Europe and North America on 14 December 1988 by the ATT Co.., British Telecom and France Telecom. TAT 8 effectively doubled the existing cable capacity across the North Atlantic and, in addition to large-scale telephone service, it transmits data and video signals [20].

TAT 8 was a major telecommunication project remarkable for the degree of international co-operation in the design, management, financing and operation of the enterprise. Ownership was vested in ATT Co. 34 per cent, BT 15.5 per cent and FT 10 per cent, the remainder being taken up by other telecommunication administrations. British Telecom with its industrial partners, notably STC, played a very significant role in planning the new system through its earlier experience in submarine cable systems and the technological support available from its research laboratories.

From the landing point at Tuckerton, New Jersey, USA the cable extends 5,800 km across the Atlantic to a branching point off the shores of the coasts of the UK and France on the European Continental Shelf; from this point the UK branch continues 540km to Widemouth, N. Devon and the French branch 330 km to Penmarch, Brittany.

The TAT 8 system provides two pairs of fibres, one pair routed from Tuckerton to Widemouth, UK and the second pair from Tuckerton to Penmarch, France. The two pairs separate at a branching repeater on the European Continental Shelf and, to keep symmetry within the cable and facilitate restoration, a pair is provided between the UK and France. The branching repeater provides for switching the fibres between the different locations for operational security reasons. Each fibre pair carries two both-way 140 Mbit/sec signals.

The main trans-Atlantic section of the cable system from the USA up to and including the branching repeater has been provided by the ATT Co., the section to the UK by STC Submarine Cables Ltd and the section to France by Submarcon. The sections provided by the various manufacturers differ in design to some degree, e.g. in the stand-by arrangements to secure high reliability. However all are based on the use of mono-mode fibre operating at about 1.3 microns wavelength.

The mechanical aspects of cable design differ between the three suppliers; each provides a home for the optical fibres along the neutral axis of the cable but uses different techniques to ensure mechanical strength and protection against water and hydrogen access under the extreme pressures encountered in the depths of the ocean. (Glass fibres are susceptible to deterioration in the presence of heavy concentrations of free hydrogen.) Heavy armouring is provided to guard against damage, e.g. by trawlers, in the shallow water sections.

A plough, designed by the BT Research Laboratories, has been used to plough in cable in shallow waters to provide additional protection against such

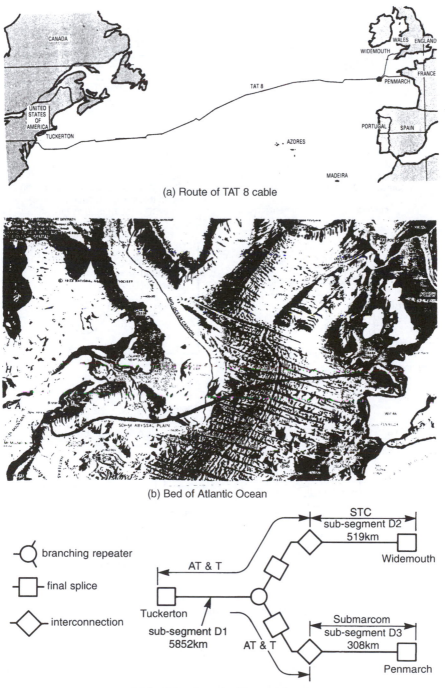

(a) Route of TAT 8 cable

(b) Bed of Atlantic Ocean

(c) Participants and cable provision plan

Figure 17.7   First trans-Atlantic optical fibre cable TAT 8

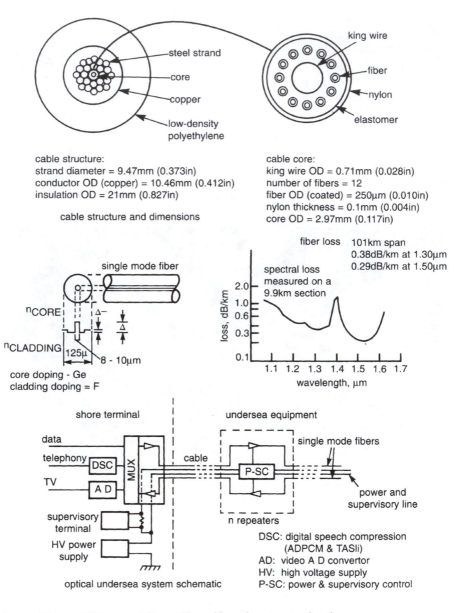

Figure 17.8   TAT 8 optical fibre cable: cable and equipment details
(Courtesy *Fiberoptics* – TAB Books)

damage. The cables were laid by the cable ships of the co-operating administrations, the UK section being laid by the BT cable Ship *Alert*. Comprehensive telemetry facilities have been provided to locate and correct faults, and facilitate the maintenance of the cable system. [19]

The success of the pioneering TAT 8 cable, the massive communication capacity it provides and the ability of digital optical fibre systems with branching repeaters to serve multiple destinations, together with the operational flexibility and economy this confers, has greatly broadened the scope for such systems.

Satellites now have a very worthy competitor for providing world-wide telecommunication services!

17.8.1 And light can now be amplified, multiplexed and switched in optical fibres

By the late 1980s some important new developments were beginning to emerge from British Telecom, ATT/Bell and other research laboratories investigating opto-electronic techniques; these included:

- the direct amplification of light in a fibre;
- multiplexing light of different wavelengths in a single fibre;
- switching light from one fibre to another;
- generating light with a high degree of coherence.

Although the amplification of light in a solid-state laser is possible, the most promising light amplifier to date is one using stimulated emission in, or with,a conventional silica-based mono-mode fibre, e.g. operating at 1.55 microns, doped with low concentrations of the rare-Earth ion erbium and pumped by light of a different wavelength, e.g. 0.98 microns. Such light amplifiers have very low inherent noise and can offer gains of 10 db or more with power outputs of up to about 1 watt. Moreover a relatively wide bandwidth is possible, permitting operation at up to at least 2.5 Gbit/sec (2,500 Mbit/sec) and opening the door to multiplexing light of different wavelengths on a single fibre with a common amplifier. [21]

Early pioneering work on light amplification in neodymium-doped glass fibre lasers operating at 1 micron was carried out by E. Snitzer and C.J. Koestler of the American Optical Company, Massachusetts in the 1960s. [22] A significant step forward was carried out by D.N. Payne and his colleagues at Southampton University in the mid-1980s; they demonstrated some 26 db gain in erbium-doped single-mode glass fibres operating at 1.5 microns, pumped at 0.65 microns. [23]

The practical implications of this advance are substantial – existing mono-mode fibre cables in the inland junction and trunk networks operating at 140 Mbit/sec can be upgraded in capacity by factors of 10 or more by replacement of repeaters; repeater spacings can be greatly increased.

A potential application also exists in future local distribution optical fibre networks to customers' premises, as a wideband power distribution amplifier handling several multiplexed signals on different light wavelengths.

The low-noise property has already found an important application in submarine optical fibre systems as a pre-amplifier to a conventional light detector, providing a valuable improvement in sensitivity and enabling a wider spacing of repeaters to be achieved.

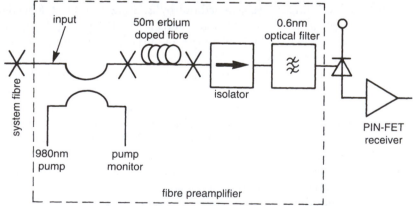

Figure 17.9   British Telecom Research Laboratories at Martlesham are awarded the
Queen's Award for Technological Achievement (opto-electronic detectors)
(1990) (left to right: Dr. Derek King, Alec Parker, and Richard Calton.)

A team at the British Telecom Research Labs, led by Dr Derek King, Alec Parker and Richard Calton, was responsible for the design and manufacture in quantity of an optical receiver for submarine systems that included a light pre-amplifier and achieved the highest sensitivity then available anywhere in the world.

The high-sensitivity optical receiver has been used in TAT 8 and numerous submarine systems in the North Sea, Mediterranean and Pacific Ocean. In recognition of this and other outstanding work on optical fibre systems BTRL were awarded the prestigious Queen's Award for Technological Achievement in 1990. [24]

17.8.2 'Blown Fibre'

Not all the developments in the optical fibre world have been on the plane of high technology – an innovation of considerable practical and economic importance has been the development of the 'blown fibre' technique for install-ing optical fibre cables in customers' premises and under pavements in the local network. The speed and low cost of fibre installation by this method have made a valuable contribution to the spread of optical networks.

In this technique bundles of three or four optical fibres, held together in a polyethelene sleeve, are blown down a small tube or 'microduct'. The microduct is a tough and flexible tube with simple 'push-fit' connector joints; using com-pressed air the bundle of optical fibres can be blown distances of 2 km or more through the microduct.

This development was carried out at the BT Research Laboratories by Mick Reeve and his colleague Steven Cassidy and led to field trials in 1984. In 1989 the Martlesham Medal was awarded to Mick Reeves. The citation for the award noted that

Blown Fibre is a fine example of applied research of the very best sort. Thanks to Mick's dedication and imagination it has been refined from his original idea into a practicable, marketable product.

References

1 C.K. Kao and G.A. Hockham, 'Dielectric Fibre Surface Waveguides for Optical Frequencies', (*Proc. IEE*, 113, No. 7, July 1966).
2 I.C. Magaziner and M. Patinkin, 'Corning Glass: The Battle to Talk with Light', Chapter 9 in *The Silent War* (Random House, New York, 1989).
3 R.D. Maurer, 'Glass Fibers for Optical Communications' (*Proc. IEEE*, 61, April 1973).
4 A.L. Schawlow and C.H. Townes. 'Infrared and Optical Masers' (*Phys. Rev.*, 112, December 1958).
5 T.H. Maiman, 'Stimulated Optical Radiation in Ruby' (*Nature*, 187, August, 1960).

6   M.B. Panish, 'Heterostructure Injection Lasers' (*Bell Lab. Rec.* 49, November 1971).

7   C.A. Burrus and R.W. Dawson, 'Small-Area High-Current Density GaAs Electro-luminescent Diodes and a Method of Operation for Improved Degradation Characteristics' (*App. Phys. Lett.* 17, August 1970).

8   J.A. Kuecken, *Fiberoptics: A Revolution in Communications* (TAB Books, 1987).

9   E.F. O'Neill (ed.), *A History of Science and Engineering in the Bell System, Transmission Technology (1925–1975)*, Chapter 20, 'III Early Optical System Work', (ATT Co./Bell Laboratories, 1985).

10 T. Hey and P. Walters, *The Quantum Universe* (Cambridge University Press, 1987).

11 K.J. Beales, J.E. Midwinter, G.R. Newns and C.R. Day, 'Materials and Fibres for Optical Transmission Systems' (*POEJl*, 67, July 1974).

12 R.B. Dyott, 'The Optical Fibre as a Transmission Line', (*POEEJ*, 67, October 1974).

13 D.H. Newman, 'Sources for Optical Fibre Transmission Systems' (*POEEJ*, 67, January 1975).

14 M.R. Mathew and D.R. Smith, 'Photo-Diodes for Optical Fibre Transmission Systems', (*POEEJ*, 67, January 1974).

15 'Martlesham Medal for Pioneer (Dr. M.M. Faktor) in Opto-Electronic Technology', (*Brit. Telecom. Eng.*, 6, April 1987).

16 'Martlesham Medal for Optical-Fibre Pioneers (Dr. G. Newns and Dr. K.J. Beales), (*Brit. Telecom Eng.*, 1, October 1982).

17 R.C. Hooper, 'The Development of Single-Mode Optical Fibre Transmission Systems at BTRL' (*Brit. Telecom Eng.*, 4, July 1985).

18 '565 Mbit/sec Optical Fibre Link' (*Brit. Telecom Eng.*, 5, April 1986).

19 'Optical-Fibre Submarine Cable Systems', (*Jl. British Telecom Engineers*, 5, Pt.2, July 1986).

20 'TAT Goes Live', (*Brit. Telecom Eng.*, 7, January 1989).

21 *British Telecom Technical Review*, (Autumn 1989 and Spring, Summer 1990).

22 C.J. Koestler and E. Snitzer, 'Amplification in a Fiber Laser' (*JL. Applied Optics*, Vol.3, No.10, October, 1964).

23 R.J. Mears, L. Reekie, I.M. Jauncy and D.N. Payne, *High-gain Rare-Earth-Doped Fibre Amplifier at 1.54 Micron*, (Dept. Electronics and Information Eng., Southampton University, 1987).

24 'Labs Win Fourth Queen's Award', (*Brit Telecom Technical Review*, Summer 1990).

Further Reading

Jeff Hecht, *City of Light – the Story of Fiber Optics* (Oxford University Press, 1999).

Chapter 18

Inventors of the visual telecommunication systems

The invention and development of the telephone enabled the simplest and most direct form of human communication to be achieved, that is by voice and ear, but left unused another human sense organ – the eye. There was clearly scope for a variety of modes of visual communication, ranging from the transmission of hand- or type-written documents, sketches and photographs (facsimile), to live video conferencing between pairs or groups of people (tele-conferencing), and teletext (the presentation of alpha-numeric and other information on television receivers or video display units).

18.1 Facsimile transmission[1]

The early history of facsimile transmission goes back to 1842 when a Scottish inventor, Alexander Bain, developed a pendulum system for synchronising a master-slave system of clocks. He realised that if the master pendulum traced a series of electric contacts, e.g. corresponding to letters of the alphabet, a slave pendulum with a stylus carrying an electric current could reproduce these signals on chemically-impregnated paper – this he called 'an electro-chemical recording telegraph', for which he was granted a British Patent No. 9745 in 1843.

The first commercial system was established in France in 1865, linking Paris with several cities and using a modification of Bain's device patented by an Italian, Giovanni Caselli.

Bain's system was later superseded by electro-mechanical printing telegraph systems such as that invented by David Hughes in 1854 (see Chapter 3). Nevertheless, by introducing the concepts of synchronisation and the creation of an image from a varying electric current, Bain had laid the basic foundations of modern facsimile systems.

A major step forward in the evolution of facsimile systems was made in 1850

---

[1] This account is based on Daniel Costigan's book. [1]

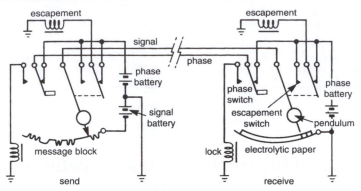

(a) Alexander Bain's 'Elector-chemical recording telegraph' (1843)

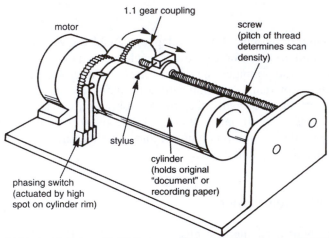

(b) Frederick Bakewell's 'Cylinder and screw' scanning system (c.1850)

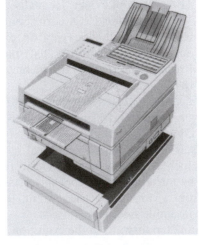

(c) Modern facsimile machine for
whole newspaper pages (Muirhead)

Figure 18.1   Facsimile systems (From D.M. Costigan's Electronic Delivery of Documents
and Graphics)

by the Englishman, Frederick Bakewell, who demonstrated the 'cylinder-and-screw' arrangement that replaced Bain's pendulum and is still used in some facsimile systems today.

The next important step was the use of the light-sensitive element selenium, first suggested by Shelford Bidwell in England, as a means for converting a scanned image into a varying electric current – a technique which opened the door to 'half-tone' transmission, i.e. of intermediate tones between white and black.

The first practical photo-electric facsimile system was demonstrated in 1902 by the German inventor, Dr Arthur Korn. Commercial use of this system began in Germany in 1907, and by 1910 Berlin, Paris and London were linked by facsimile transmission over the telephone network. By 1922 Korn's system had been used to transmit by radio a photograph of Pope Pius XI from Rome to Maine, USA, the received picture being published the same day in the *New York World* newspaper. Thus began what was to become one of the most valuable uses of facsimile transmission – by the Press, initially for individual photographs of up-to-the-minute news, but eventually for whole pages of newspapers from news-gathering offices in cities to remote presses.

Following Korn's successful demonstrations, facsimile transmission systems – the term is most commonly shortened to 'fax' – began to be developed commercially by the Marconi Co. and by Muirhead in England, Telefunken in Germany and Belin in France. In the USA the ATT Co., supported by pioneering research in the Bell Laboratories, [2] the Radio Corporation of America and the Western Union Telegraph Co, set about developing picture transmission systems, initially for Press use.

One of the technically most interesting facsimile systems, developed in England in 1920, two years before Korn's trans-Atlantic radio triumph, by H.G. Bartholomew and M.L. MacFarlane and called the 'Bartlane' system, enabled picture transmissions to be made over the Trans-Atlantic telegraph cable (see Chapter 3). It converted picture elements to groups of perforations in paper tape in a manner anticipating pulse-code modulation PCM digital systems. The perforated paper tape was used to transmit on-off signals over the cable to the receiving end where each group of coded signals re-created a picture element. Clearly, because of the restricted bandwidth available on the cable, this was a very slow process.

In addition to the Press usage of facsimile, RCA pioneered its use for the transmission by radio of weather maps to ships at sea, from which developed a major usage in weather forecasting.

Some of the early facsimile workers were seized with the idea of using radio broadcasting transmitters to transmit newspapers, either during the night hours when the transmitters were not in use for sound broadcasting, or on an amplitude or frequency modulated tone higher in frequency than the sound channel. One fax system called the 'Fultograph' after its Austrian inventor Captain Otho Fulton was operated experimentally for a few years beginning in 1928, in at least four European cities, including Berlin.

Several fax newspaper systems were tried out on a semi-commercial basis in the USA during the 1930s and 1940s, and by 1948 the US Federal Communications Commission had authorised commercial fax broadcasting. But by then television broadcasting had made its debut and was commanding public attention; fax broadcasting virtually ceased but fax continued to develop in its more utilitarian roles for Press, weather map distribution and business users.

In the early 1930s a number of developments took place that later significantly improved the quality, speed and convenience of facsimile transmission. One was the all-electronic flying-spot scanner invented by Manfred Von Ardenne in Germany, and the other a modulated ink-jet recording method developed by C.W. Hansell of RCA. John Logie Baird in England (see Chapter 9) was investigating in 1944 the possibility of using his television scanning techniques for high-speed facsimile transmission in wide-band radio channels. Finch in America demonstrated a system of colour fax, a high-speed cathode-ray tube scanner and, in 1938, the simultaneous transmission of fax and high-quality sound on FM VHF broadcasting transmitters.

The development of television spurred on other electronic device developments that found application in facsimile systems, including the photo-multipliers pioneered by Zworykin and Farnsworth. The facsimile art itself continued to develop from the 1950s onwards, notable advances being electrostatic recording (Xerography, RCA's 'Electrofax'), improved scanning by lasers and Bell Laboratories charge-coupled devices (CCDs).

Transmission was speeded by techniques that reduced redundancy, e.g. by fast scanning over white spaces between letters and words, evolved by Kretzmer, Hochman and others. By the late 1970s the microchip and the micro-processor had become available – see Chapter 12; this simplified and improved the operation of fax equipment, made it more flexible and at the same time contributed substantially to reliability and economy.

These developments, which enabled the large-scale manufacture of compact, low-cost fax equipment, opened the door to a vast new market for business and personal use. By the 1980s facsimile was an accepted part of the armoury of information technology that is transforming the business world.

Vitally important for the growth of the facsimile transmission market was international standardisation of the technical characteristics and operating procedures through the International Telephone and Telegraph Consultative Committee (CCITT), initially directed to the use of 3 kHz bandwidth telephone channels. At first, in the 1960s, it took six minutes to transmit an A4 page (Group 1 standard). By 1976 the Group 2 standard had reduced this to three minutes and improved the quality with a scanning density of 100 lines per inch.

Further developments followed with Group 3 machines, which use digital encoding and transmit an A4 page in less than one minute over a telephone channel, with high-quality reproduction of text and diagrams.

A new generation of facsimile machines – Group 4 standard – is set to surpass the Group 3 machines as the Integrated Services Digital Network (ISDN) becomes established on a global basis. By exploiting the 64 kbit/sec capability of

Figure 18.2   British Telecom 'Video-phone' (1994)

the ISDN the transmission time of an A4 page is reduced to less than 30 seconds, with faster connection set-up time between sender and receiver.

The provision of hard copy of documents by facsimile techniques further enhances the visual communication capabilities of the ISDN which include person-to-person tele-conferencing and VDU document display via the Internet (see Chapter 20).

18.2 Television conferencing

From the 1950s onwards an increasing effort was being put into research aimed at providing a live visual element to person-to-person and group-to-group communication at a distance, with the objective of making such communication as close to direct contact, in the same room, as was practically and economically possible.

18.2.1 The analogue video-phone

The first efforts towards this objective fell a long way short of the ideal – they took the form of the 'Picturephone' development by the Bell Telephone

Laboratories in the USA [3] and a similar development in the UK, the British Post Office 'Viewphone'. Both offered the telephone user a head and shoulders picture of the distant telephone caller and limited capability for displaying alpha-numeric data. The 'Picturephone' used a 251-line, 30 frames/sec picture requiring a 1 MHz video bandwidth, the 'Viewphone' offered a 319-line, 25 frames/sec picture in a 1 MHz video bandwidth. Even with this restricted bandwidth the links to the customers premises and for inter-city transmission were complex and costly – the 'Picturephone' required three cable pairs to each customer's premises, one pair for voice and two pairs for the go and return video signals.

Even when the video signal was encoded to the lowest practicable rate of 6.3 Mbit/sec for inter-city transmission, only three long-distance circuits could be provided in the 20 MHz baseband of microwave radio relay systems. Extensive ATT Co./Bell System field trials of 'Picturephone' service were carried out in 1965/6 between New York and Chicago but failed to attract customers on a sufficient scale to justify further development of the service.

But by the early 1990s prospects for the 'video-phone' had improved dramatically. The ongoing development of digital and coding methods had opened the door to new techniques for drastically reducing the bit-rate by the 'discrete cosine transform (DCT)'.

This made it possible to transmit acceptable quality colour video signals, albeit at the relatively low picture rate of 8 per second, over customers' conventional analogue telephone lines. By 1994 both ATT and British Telecom were offering video-phone service based on the new digital technology, although at different bit rates (ATT at 19.2 kbit/sec, BT at 14.4 kbit/sec).

18.2.2 The higher definition digital video-phone

With the advent of ISDN a higher definition video-phone becomes possible. In a British Telecom development, a 64 kbit/sec channel in the ISDN is used, with advanced coding of the video signal and a picture repitition rate of 15 pictures per second, to provide picture quality approaching that of normal broadcast television. The picture can be displayed on a conventional television monitor or as a 'window' on a computer VDU; the latter arrangement has value for sales and shopping services as well as providing person-to-person visual contact.

18.2.3 Group-to-group tele-conferencing

Television conferencing or tele-conferencing can provide a service of great value for business users and large organisations with dispersed office or factory locations. The ultimate objective is to achieve electronically the same ease of communication between groups of people separated by distance as if they were in the same room. The value of tele-conferencing to the users increases with distance because the costs of travel and staff time are themselves distance related.

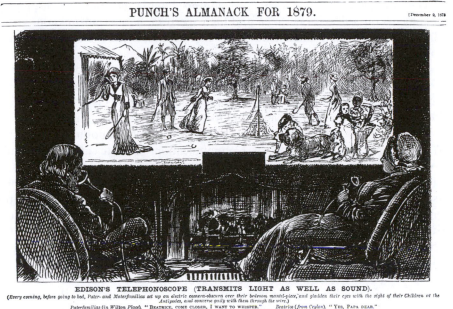

EDISON'S TELEPHONOSCOPE (TRANSMITS LIGHT AS WELL AS SOUND).

(Every evening, before going to bed, Pater- and Materfamilias set up an electric camera-obscura over their bedroom mantel-piece, and gladden their eyes with the sight of their Children at the Antipodes, and converse gaily with them through the wire.)

Paterfamilias (in Wilton Place). "BEATRICE, COME CLOSER, I WANT TO WHISPER." *Beatrice (from Ceylon).* "YES, PAPA DEAR."
Paterfamilias. "WHO IS THAT CHARMING YOUNG LADY PLAYING ON CHARLIE'S SIDE?"
Beatrice. "SHE'S JUST COME OVER FROM ENGLAND, PAPA. I'LL INTRODUCE YOU TO HER AS SOON AS THE GAME'S OVER?"

Figure 18.3   Mr Punch foresees conference television 100 years ago

It is remarkable that this cartoon in *Punch* magazine for December 1879 foresaw not only UK–Australia Conference Television but large screen projection in a format not dissimilar from today's high-definition TV!

It is important, therefore, to minimise the costs of long-distance transmission, as well as the link to the users' premises, for a tele-conferencing service to be economically attractive.

For maximum effectiveness of communication a group-to-group television conference facility requires a higher definition and larger picture than is available in the 'video-phone' services; a natural first choice in exploring the possibilities was the television broadcasting standards of 525-lines in the USA and 625-lines in the UK and Europe for which cameras, television receivers and inter-city transmission facilities are readily available.

In the early 1970s the British Post Office Research Department carried out experimental field trials of a television conference service using modified broadcast standard television equipment, enabling groups of up to about ten people to communicate visually as well as aurally. The research studies included means for video bandwidth reduction by optimising picture format for the conference service and reducing redundancy in the picture signal; by so doing two conference video channels could be provided in a broadcast standard inter-city video link.

In 1972 the BPO opened its first 'Confravision' service, with studios in London and four other cities in the UK for hire by business and other users, claimed as 'a world-first' purpose-built conference television service. [4]

The conference studios were also equipped with facilities for displaying documents on a television screen and high-speed facsimile transmission, virtually enabling copies of documents to be 'handed across the table' as in a direct conference situation.

The long-distance capabilities of 'Confravision' were further demonstrated in 1979 with the setting up of Europe's first conference by satellite between London and the *'Telecom '79 International Conference* in Geneva. [5] And in 1984 the Ford Motor Company and British Telecom International set up a Confravision link between studios in its Essex plant and another in Cologne, W. Germany, via the European Communication Satellite ECS1, for daily consultations between its staff at the two centres. [6]

In the 1980s the television broadcasters began to use television conferencing as a ready and effective means for conducting interviews, generally on a 'one-to-one' basis, and often on a world-wide scale via comunication satellites. There was also the occasional use of multiple screen, or split-screen, techniques for conducting discussions involving participants in three or more distant locations simultaneously.

Up to the late 1990s the use of tele-conferencing by business and other organisations has been on a somewhat limited scale, in spite of the manifest advantages of immediacy, convenience and time saving it can offer.

Developments that are opening the door to larger scale use of tele-conferencing include:

* lower costs for local area, inter-city and international transmission;
* the ability to 'dial up' a video link as readily as a telephone connection;
* the facility for displaying simultaneously participants in three or more different locations.

18.2.4 Tele-conferencing and the integrated services digital network (ISDN)

The development of an integrated services digital network (ISDN), see Chapter 19, will enable video conferencing to be achieved on a world-wide scale at viable cost using transmission and switching facilities common to telephone, facsimile and other services. A European collaborative project 'MIAS' (Multi-point Multi-media Conference System) enables groups of users at different locations to see and speak to one another, send facsimile documents, exchange data files and examine still-image presentations. [7]

Two tele-conferencing modes using ISDN are offered:

* the 'video-phone', via a 64 kbit/sec digital channel, providing picture quality adequate for person-to-person communication and telephone speech;
* a 'video-conferencing' mode at 2 Mbit/sec with higher-quality pictures suitable for group-to-group communication.

This development has been made possible by advanced designs of 'codecs' – a 'codec' (signal coder/decoder) converts analogue signals for speech, facsimile or pictures into digital form for transmission over a digital path and reconverts them at the receiving end. By using sophisticated coding algorithms the bit-rate needed for desired speech or picture quality at the receiving end can be substantially reduced. It is the development of this technique, in which members of the BT Martlesham Laboratories played a prominent part, that enables the video-phone to overcome the practical and economic limitations of the 'Picture-phone' and 'Viewphone' referred to earlier in this chapter. Furthermore the ISDN also provides the key to readily dialling up video-phone and tele-conferencing connections. [7]

18.2.5 The future of tele-conferencing

Optical fibre systems, with their wide available bandwidth and low losses (see Chapter 17), leading to lower costs per Mbit/sec per kilometre, have very important roles to play in both long-distance transmission and the local network. There are roles too for millimetric radio systems, both in the local network and for providing additional capacity in communication satellites. The British Telecom development of digital transmission Asymetric Digital Subscribers Loop (ADSL) on the copper wire pairs linking customers' telephones to local telephone exchanges enables good quality video transmission to be made via the existing local network.

There is scope for improving the communication effectiveness of group-to-group television conferencing from the user viewpoint by larger screens with higher definition – to which current developments in 1,250-line television broadcast systems using more sophisticated coding may point the way. The field of 3-dimensional presentation capable of a 'virtual reality' display is as yet inadequately explored. A single camera and microphone can only represent the eyes and ears of one member of a spread-out group of people; realism requires that each participant sees and hears the distant group from his distinctive viewpoint.

Bearing in mind the very substantial benefits that effective and economic tele-conferencing could bring to the business world, educational, political, social and other organisations, and the environmental advantages of reducing travel, it is clear that there is ample scope in this field for future innovators and invention!

18.3 Viewdata (Prestel) and Teletext (Ceefax and Oracle)

Whilst the aural presentation of recorded information, e.g. in the telephone system for instructional purposes, has a certain personal feel, visual presentation has significant advantages:

- information can be presented at a rate closely related to the recipient's ability to absorb it, since it can be held on the screen for as long as is needed;
- it can include graphic information;

- it can be more readily made 'inter-active', i.e. responsive to user requirements;
- it lends itself to permanent record form, e.g. on the printed page.

These considerations led in the 1970s to an active, but initially independent, investigation of the possibilities of including a visual information service in the British Post Office telecommunication system, and by the BBC and the Independent Broadcasting Authority in their television broadcasting systems.

18.3.1 Viewdata (Prestel)

The starting point in the BPO Viewdata development was a memorandum in 1970 from the then Managing Director, Edward Fennessy, to Heads of Departments seeking means for making more effective, and profitable, use of the telephone system – pointing out that the customers' telephone and local network were used on average for less than one per cent of the time. It was noted in the PO Research Department that, at the time, there were some 20 million telephones in use in the UK, and about the same number of television receivers, most of them in the same locations as the telephones. This immediately suggested the possibility of providing a visual information service by displaying in alpha-numeric form on the television receiver screen information derived from a databank at the exchange over the telephone line.

The evolution of this idea into a prototype 'Viewdata' system was largely the work of Sam Fedida, supported by Keith Clarke and other colleagues in the PO Research Department. A laboratory 'breadboard' model was first demonstrated at the Research Centre at Martlesham in 1971, followed by a full-scale demonstration in 1974 of a Viewdata system working over exchange telephone lines and providing access to a wide range of information sources.

The basic elements of the Viewdata system comprised:

- a television receiver with a Viewdata adaptor or decoder, either built in or added to an existing set;
- a keypad, similar to a pocket computer, with the numerals 0 to 9, and * # symbols;
- a database and a computer for control of the system, located at a telephone exchange.

The key pad gives the user interactive control of the system via a simple operating protocol, enabling him to call up first a 'menu' of the types of information available and then by successive stages of selection any one of 100,000 or more pages of information within seconds.

The control signals are sent to the exchange computer over the telephone line at a rate of 75 bits/sec; information signals are received from the exchange datbank at 1,200 bits/sec. The decoder stores a frame (page) of information and displays it continuously in alpha-numeric form on the television receiver

Figure 18.4   BT 'split-screen' confravision studio

screen. The display can make full use of the colour capability of the television receiver.

Later developments made possible the display of detailed graphics and photographic material.

The range of information sources that can be accessed are virtually unlimited and include remote databanks in addition to those at the local exchange. Typical sources are:

- telephone directories and yellow pages;
- commodity/share prices;
- items for sale;
- travel timetables;
- leisure activities;
- what's on;
- news.

The services that could be provided include:

- banking and cash transfers;
- shopping from home;
- message (telegram) transmission;
- booking facilities.

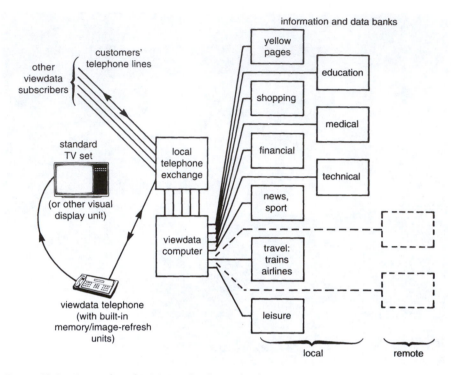

Figure 18.5   Principles of PO 'Viewdata' visual information service (1974)
(This later became British Telecom 'Prestel' service)

Following the initial demonstration in 1974 it rapidly became apparent that Viewdata was a development of major importance, and a new PO Headquarters Department with Dr Alex Reid as Director was set up to co-ordinate development and evolve business and marketing strategies. An important part of this work was to create a wide and useful database by attracting a large number of information suppliers offering material of interest and value to users.

The term 'Prestel' was adopted to describe the commercial form of 'Viewdata' – the latter term then being used generally to describe visual information services, including the broadcast version 'Teletext'.

With the handover from Post Office Telecommunications to British Telecom the further development of Prestel was continued for a time, with an increasing range of information sources and services for users. Although Prestel is no longer available as a public service, many of the facilities it offered are now available via the Internet. [8]

From the viewpoint of the provider of information, Prestel, unlike the broadcast Teletext services, offered the possibility in principle of charging the customer for the information selected according to its value.

Prestel's virtually unlimited information capacity, fast access and capability for meeting local as well as nation-wide information needs were additional

Figure 18.6 BT 'Prestel' Viewdata equipment

Figure 18.7 Prestel can provide up-dated 'Yellow Pages' information electronically at the touch of a button

advantages. Its use of existing local and national telephone networks facilitated speedy and economic provision of the service. In many ways Prestel and its progenitor Viewdata were precursors of the Internet, as was recognised by a leading financial expert:

Few people now doubt that electronic information systems of the type pioneered by Prestel will sweep across the developed world, just as the printed word did after William Caxton's invention.

(Max Wilkinson)

For their pioneering work on Prestel British Telecom were presented with the MacRobert Gold Medal, Britain's premium engineering award, by HRH the Duke of Edinburgh in 1979, and the MacRobert Award of £25,000 was presented to Sam Fedida as the inventor of Viewdata (Prestel).

18.3.2 Teletext (Ceefax and Oracle)

In the UK the Research and Engineering Departments of the BBC and Independent Broadcasting Authority (IBA) were quick to realise the possibility of using the vertical (inter-frame) interval of the television signal waveform for transmitting alpha-numeric information that could be presented on the screens of domestic television receivers equipped with a decoder. (This interval had earlier been used to transmit specialised test signals such as the sine-squared pulse 'K' rating signals pioneered by Dr N.J. Lewis of the British Post Office Research Department in the 1950s and used for assessing the transmission quality of television links.)

The initial development of Ceefax and Oracle by the BBC and IBA proceeded independently of one another and of Prestel; however, after initial ideas had been clarified it became apparent that compatability between Ceefax and Oracle was essential, and a degree of compatability with Prestel was desirable to gain the support of the television receiver industry and wide public acceptability.[2]

IBA's development of Oracle (an acronym for Optional Reception of Announcements by Coded Line Electronics) was carried out in the Authority's Experimental and Development Department under the direction of the late Howard Steele, where the work was pioneered by Peter Hutt supported by George McKenzie and other colleagues.

A paper presented by Peter Hutt at an International Broadcasting Conference in London in September 1972 (IEE Conference Publication No. 88), on the labelling of programme sources in the complex IBA network by data inserted in the television waveform included, amongst other possible applications, the following: 'Transmission of captions to special domestic receivers. Regional news and/or weather service distribution'.

From this beginning development work proceeded rapidly and IBA began its first 'on the air' demonstrations of Oracle in April 1973.

---

[2] The author is indebted to Mr Pat Hawker for an account of the early history of Oracle/Ceefax development.

The BBC's development of Ceefax was carried out at the Kingswood Warren Research Laboratory, where S.M. Ewardson and J.P. Chambers made notable contributions. A simulated demonstration of the possibilities of Ceefax was given in the 'Tomorrow's World' television programme in October 1972.

Compatability between the Ceefax and Oracle systems was eventually reached through a joint BBC, IBA and TV Industry Committee under the auspices of the British Radio Equipment Manufacturers' Association, and regular transmissions by the BBC and IBA commenced in the mid-1970s. Similar systems began to be developed in France, Germany, the USA and Japan, and discussions took place in the Videotext Committee of the CCITT with a view to an international standard system. This was launched in 1976, based largely on the British Ceefax/Oracle systems, and designated the World System Teletext (WST) Standard. [9]

Broadcast Teletext systems offer up to two or three hundred pages of information (limited by the time available in the vertical interval of the television signal waveform). The desired page is chosen from an index page and accessed from a push-button unit by dialling a three-digit page number, with an average response time of a few seconds.

The menu offered by Teletext within the scope of the two or three hundred available pages is very wide, including weather and travel information, news, financial information and stock exhange prices, details of radio and television programmes, what's on in theatres and cinemas, holiday offers and for sale notices.

Of particular value for the deaf are the captions that can be inserted when viewing some television programmes. The Teletext service itself is free to viewers and available throughout the area covered by normal television broadcast services.

It is fitting that the BBC and the IBA were in 1983 jointly presented with the Queen's Award for Technology for their pioneering work in creating the world's first successful Teletext service.

References

1 Daniel M. Costigan, *Electronic Delivery of Documents and Graphics* (Van Nostrand Reinhold Co., 1978).
2 Horton *et al*, 'Transmission of Pictures Over Telephone Lines' (*Bell System Tech. Jl.*, April 1925).
3 E.F. O'Neill (ed.), *A History of Science and Technology in the Bell System: Transmission Technology (1925–1975)*, Chapter 22 'Services' (ATT/Bell Labs, 1985).
4 'Post Office Launches Confravision' (*PO Elec. Eng.Jl.*, 64, January 1972).
5 'Confravision by Satellite' (*PO Elec. Eng. Jl.*, 72, January 1980).
6 'International Video Conference Service' (*Brit. Tel. Eng.*, 3, October 1984).
7 'The Codec Story' (British Telecom, *Competitive Edge*, Issue 4, May 1991).
8 Dr A. Reid (ed.), *Prestel 1980* (British Telecom 1980).

9 G.O. Crowther, *U.K. Teletext – Its Evolution and Relation to Other Standards* (IEE International Conference on 'The History of Television', Conference Publication No. 271, November 1986).

Chapter 19

Information technology and services: data communication

To an increasing extent, social, economic and political activities in the modern world are becoming dependent on information technology (IT) – the means by which information in the form of data is processed, stored, accessed and transmitted. The term 'data' in this context includes numeric, voice, music, facsimile, still and moving picture (video) signals.

The advances in telecommunications outlined in earlier chapters have removed the barrier of distance and reduced the costs of transmission so that users and databanks can be remote from one another. The microchip has enabled vast amounts of data to be stored, rapidly processed and accessed at costs that continue to fall. It is now clear that information technology will have an increasing influence on the way people live and work, the quality of their lives and the economic prosperity of nations. By reducing the need to travel to communicate the environment itself may benefit. [1] [2]

19.1 The impact of telecommunications and information technology (IT) on the world scene

The advances in telecommunication and IT made after the 1960s have already had a dramatic and significant impact on the world scene. It is appropriate to recall some instances that have caught the imaginations of, and are stored in the memories of, millions of people:

- the telephone 'hot-line' between Presidents Kennedy and Krushchev that staved off World War III in the Cuban missile crisis of 1962;
- the television pictures transmitted by the Telstar satellite that showed, in agonising clarity, the assassination of President Kennedy in 1963;
- the virtually instantaneous transmission by electronic means of millions of dollars that facilitated the over-rapid computer-controlled buying and selling of shares that precipitated the world stock-market crash of 1987;

- the live television pictures of the first Moon landing and the conversation of President Nixon with the astronauts Neil Armstong and Buzz Aldrin in 1969;
- the detailed pictures of the furthermost planets of the solar system made possible by picture-storage, slow-scan IT techniques;
- the live action pictures and commentaries by news correspondents, via satellites and compact microwave transmitting equipment, from war and disaster areas world-wide.

And it may well be that the existence of intelligent life on remote planets of our star galaxy will be revealed by the search of the radio spectrum now being conducted using large microwave aerials and IT-based signal recognition techniques. But perhaps one of the most exciting of uses of IT has been to process the pictures of star galaxies at the most distant edge of the universe, revealing what they were like at the time of the 'Big Bang', 15 billion years ago.

However, it has to be admitted that there is a possible 'down-side' to the uses of IT – exemplified by the activities of 'hackers' who have used their knowledge of IT and system organisation to gain illegal access to the databanks of private individuals, large firms, political organisations and even the 'War Games' databank of the US military at the Pentagon, Washington! And there are dangers associated with the spread of pornography and terrorist activities that have to be guarded against.

19.2 Team development of IT

The ongoing development of the telecommunications applications of IT has been made possible largely by advances in device technology, such as the microchip, and software/operating system techniques, common to the computer field. In recent years these developments have tended to be carried out by team work in the laboratories of large telecommunication organisations such as British Telecom and ATT Co./Bell, and computer firms such as IBM, rather than by readily identifiable individuals in the first stages of development, as described in earlier chapters of this book. Nevertheless it should be remembered that Microsoft itself began from the enterprise and imagination of two young men! Furthermore, the need for international standardisation of technical parameters to facilitate the setting up of world-wide networks has meant that this aspect of development has to be carried out via international committees, for example of the International Telecommunication Union.

19.3 IT and telecommunication services

It is appropriate at this stage to consider in more detail the nature of 'information technology' and 'information'.

Information technology is primarily concerned with the manipulation of data. For present purposes 'data' is taken to refer to spoken or written words, letters, punctuation marks, numbers and other symbols, graphic and photographic material, and video signals, in analogue or digital form.

'Information' is an assembly of knowledge, facts, ideas, instructions and the like which can be expressed by data, and which is meaningful to, but not necessarily new to, a recipient. Claude Shannon defined the basic unit of information as the 'bit', corresponding to an on-off or yes-no signal (see Chapter 13).

The elements of an information technology system comprise 'computers', in which information is stored as data and processed as determined by a control program ('software'), and 'telecommunication networks', with terminal equipment for the insertion, selection and display of data, providing data transmission and switching systems.

An important feature of developments since the 1960s has been the convergence of computer and telecommunication technology, exemplified for example by the common use of micro-processors, stored-program control and digital methods.

The visual display units (VDUs) that are a feature of computer systems owe much to the development of the television picture tube, while the computer printer is related to the facsimile machine. Thus it might be said that the innovators and pioneers of key advances in telecommunications have benefited the development of computers, and vice versa. And both groups have contributed to the creation of information technology.

The 'telecommunication services' available under the generic title of 'information technology' is wide and expanding; they include audio conferencing, teleprinter transmission, data transmission, facsimile transmission, television conferencing, mobile radio and electronic mail (e-mail).

19.3.1 Audio conferencing

Audio conferencing facilities are available enabling three or more ordinary telephone users at different locations to be linked on a dial-up basis via the public switched telephone network or via private circuits. A more extensive audio conferencing facility enables larger goups, e.g. in halls, to confer via loudspeakers, microphones and voice-switching equipment. The ISDN enables audio conferencing to be provided more quickly, with greater economy and better speech quality than is possible over 3 kHz bandwidth telephone channels; advanced codec design for analogue speech signal to digital conversion enables high-fidelity speech transmission to be achieved over 64 kbit/sec digital channels.

19.3.2 Teleprinter transmission (Telex)

This provided an automatic electric typewriter public switched service with transmission by digital signals using internationally agreed (CCITT) codes and alpha-numeric alphabets, typically at a speed of 50 bits/sec. The service, initially based on relatively slow electro-mechanical machines, has largely been super-seded by faster, quieter electronic systems using facsimile and visual display (VDU) techniques, and by the electronic mail (e-mail) service.

19.3.3 Data transmission

Data transmission at speeds of 9,600 bits/sec can be made over the voice circuits of the public switched telephone network, via modems that convert the digital signals to keyed audio tones. Modern modems operate at 56 kbit/sec higher speeds, typically 64 kbit/sec or multiples of this up to 2 Mbit/sec, are available over the ISDN, the faster speeds being valuable for the transfer of large capacity databanks and the provision of high definition facsimile, e.g. for whole-page newspaper transmission.

19.3.4 Facsimile transmission (Chapter 18, section 18.1)

Facsimile systems provide good-quality transmission of hand-written or type-script material and diagrams over telephone circuits, with a transmission time of less than one minute for an A4 sheet. Faster transmission and higher quality, including the transmission of half-tone photographic and colour material, is possible using more sophisticated coding techniques over the ISDN.

19.3.5 Television conferencing (Chapter 18, section 18.2)

Group-to-group television conferencing is available over inter-city and inter-national cable, microwave and satellite television links and will become increasingly economic as more advanced coding techniques reduce the video band-width requirements. The use of optical fibre systems, with their wide available bandwidth, will further reduce transmission costs. The ISDN will provide person-to-person 'video-phone' conferencing, as well as group-to-group higher-definition video conferencing.

19.3.6 Mobile (cellular) radio (Chapter 21)

The last decade has seen a massive development in the UK, Europe, America and Japan of mobile radio-telephone systems using hand-held portable equipment operating with frequencies in the UHF band via an extensive over-lapping 'cellular' network of base radio stations.

   These provide access to national telephone systems using technical standards and procedures standardised by the International Telecommunication Union.

Current developments enable access to the Internet and the small-screen display of messages and other visual material.

19.3.7 Electronic mail (e-mail)

Beginning in the early 1990s there has been a remarkable world-wide development of 'electronic mail' (e-mail) – a person-to-person fast communication technique via keyboard and visual display, with paper copy. E-mail is character based, although digitally encoded images may also be transmitted. It uses alpha-numeric presentation, with digital transmission using the ASCII code over conventional telephone circuits. As such it has features in common with the earlier 'telegram' and 'Telex' (teleprinter) services but provides much speedier and more flexible service. E-mail provides convenient two-way communication, with messages stored in remote server 'mailboxes' until required. It also offers a 'bulletin' or notice board containing information of permanent or semi-permanent nature, relevant to a given organisation or group. And above all it is a relatively low-cost service, since it can be provided over ordinary telephone circuits. However, it has to be remembered that other people have access to the user's computer and messages are thus open to interception for purposes which may not be legitimate. Furthermore the connection of a computer into an unlimited network may make it vulnerable to the injection of the computer 'virus', i.e. illegal instructions that may damage the computer operating system and destroy or garble information.

The success of e-mail is largely due to the creation of the Internet, which has enabled national and international telecommunication systems to be linked via computer-based service providers (see Chapter 20).

19.4 The convergence of information and computing technology

The development of telecommunication-based information processing and databank accessing systems such as those outlined above has been made possible by the parallel development of computing technology. The latter includes the following basic and highly significant developments:

- the 'ASCII code' (American Standard Code for Information Exchange), in which the lower and upper case letters and the numerals, the punctuation marks and spacing between letters, are each allocated a number up to 100 which also identifies a key on the computer keyboard;
- the 'Computer Disc' (CD) which enables many millions of bits of information to be recorded on spiral tracks on a disc a few centimetres wide and read either magnetically or by a fine laser beam; the computer operating system can be arranged precisely to locate and record, or read, on the disc selected areas of information;

- the system known as 'CD-ROM' or read-only memory and the concept of computer control by 'software', the digitally encoded instructions, recorded on disc or tape, which instruct the computer to carry out prescribed functions and operations;
- the specific software concept known as 'Microsoft DOS', where DOS refers to Disc Operating System (Microsoft is a US company trademark); the origin of Microsoft DOS, which is used for computing world-wide, has become an essential element in information technology and has resulted in a billion-dollar industry in the USA;
- linked with 'Microsoft' is the important 'Windows' concept which enables chosen symbols (icons), alpha-numeric information, still or video pictures to be placed as required in selected areas of a VDU screen;
- a screen cursor, controlled by a manually operated control or 'mouse', enables an icon denoting and activating a chosen function to be selected at will; these concepts are particularly significant for the 'multi-media' development of information services;
- the advent of the low-cost personal computer/word processor, made possible by the falling costs of microchips, which has not only replaced the large and costly ' main-frame' computer for many purposes but has enabled information technology to reach into the home as well as the office.

In the UK Alan M. Sugar and in the USA Ed Roberts have made valuable contributions by initiating the development of low-cost 'personal computers'.

19.5 The origins of 'Microsoft DOS'

Much of the credit for the creation of Microsoft DOS must go to the vision and enterprise of a young American, Bill Gates, who with college friend, Paul Allen, realised in 1975 that the then existing commercially available computers lacked the flexibility to handle personal computing/word processing. Together, in Seattle USA, they produced the first microcomputer software that helped to launch the PC revolution [3].

By 1979 Gates and Allen had licensed their software to other US computer manufacturers; rather late in the day the giant computer firm IBM realised that its future in the PC field lay in the hands of Gates and Allen and a small company called Microsoft holding basic patents in PC computer software.

From these beginnings, and with a computer operating system called *Q.DOS* bought from another Seattle company and translated into Microsoft by Gates and Allen, in 1981 IBM launched a 'personal computer' that became familiar to computer users all over the world under the acronym MS-DOS (Microsoft Disc Operating System). Other US computer firms entered the battle, and Apple developed a personal computer with an improved operating system offering 'icons' and 'windows'.

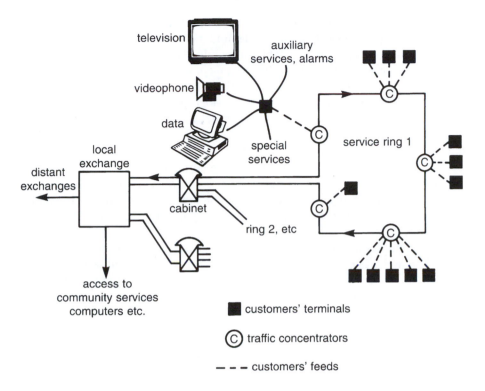

Figure 19.1   Prototype broad-band local distribution network using a digital coaxial cable ring main

(British Post Office Research Laboratories, 1968–1972)

By progressive innovation and enhancement of its original patents Microsoft has maintained a dominant position in the computer software field – a position which, however, has recently been challenged by others via the US Department of Trade and Industry and the US Justice Department as possibly harmful to competition in the computer industry.

19.6 The integration of IT and home/business communication services

19.6.1 Home services

As the range of information and entertainment services available to customers in their homes expanded it became logical to seek ways of providing these on a common local distribution network since this could reduce the costs of provision and be more convenient to the users, as compared with separate provision of the various services.

One such forward-looking study was made by the 'Subscribers Apparatus'

Group at the British Post Office Research Department, Dollis Hill in the late l960s and a prototype laboratory demonstration model was made. This embodied a broadband coaxial cable ring main providing a digital highway to which customers' premises were linked in conveniently located groups over feeder cables or short-range microwave links, via traffic concentrators.

The services then envisaged included video-telephones, data VDUs and access to community databanks and computers, broadcast television and auxiliary services such as fire alarms. The project, which was recorded in a colour film *Telecommunications 1990*, was perhaps ahead of its time. Nevertheless, the advent of the microchip, optical fibre cables and the spread of digital transmission and switching throughout the trunk network have now brought it nearer practical reality. To these entertainment services might be added on-demand access to an electronic 'video library' service discussed later in this chapter.

Current studies of 'the home of the future' envisage a wide range of functions that IT could play, ranging from voice and visual communication, information access, entertainment services, shopping and banking from home, home security alarms, and 'working from home'. However, at the present time political ideology and the claimed merits of competition appear to inhibit

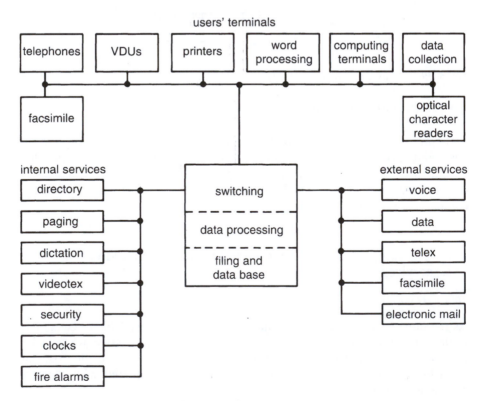

Figure 19.2   The integration of business information services

the full integration of entertainment and communication services in the home.

19.6.2 Business services

The impact of information and computer technology in the business field, in banking and in government at national and regional levels is substantial and expanding. In the business field it provides immediate communication between individuals and groups, rapid access to data files on such diverse topics as sales, stock piles, cash flow, personnel, group performance, and stock-exchange share prices, and provides a mechanism for large and small cash transactions and transfers. Within a business it can provide a range of services including directory enquiries, paging, dictation, videotex (display of information on visual display units), security and alarm facilities. By voice, facsimile, telex, electronic mail and audio/video conferencing services these facilities can be extended to other locations, e.g. from business headquarters to other offices, factories and sales units, without regard to distance.

One commentator has noted:

> ... the power of information technology to supply managers with accurate information on the day-to-day performance of groups ... and the ability of information technology to condense information into readily understandable charts and figures ... allows chief executives to control larger teams than was possible using pre-computer management methods.

However, there is a note of caution in the familiar phrase from the computer field 'garbage in, garbage out', i.e. the output from an information system can be no more reliable than that of the data supplied to it.

19.7 Data communication

Data communication involves a logical structure that enables data terminals (data sources/receivers) to be interconnected via transmission networks involving one or more computers or switching centres. The general objective is to ensure that message information – data – reaches its intended destination(s) without error and with a minimum time occupancy of the transmission network. As the numbers of terminals and computers increase, so does the complexity of network organisation if the objective of efficient use of transmission paths is to be maintained. The reader is referred to specialised texts for a detailed account; for present purposes the following modes of operation of data networks may be distinguished: [3] [4]

19.7.1 Circuit switching

This is the mode generally used in telephone systems where calls are 'dialled up' and switched at one or more exchanges via electro-mechanical or electronic switches, the transmission path being held 'open' for the duration of the call, irrespective of whether message information is passing or not. From the latter point of view it is not a highly efficient use of the network; however, the rate of data transmission is limited only by the bandwidth of the transmission path and, since no 'store and forward' operation is involved, there is no traffic congestion.

19.7.2 Message switching

Message switching is particularly adapted to multiple node networks, i.e. involving several computers and transmission paths. Each computer node has processing capacity that enables it to:

- interpret an address at the head of each message;
- temporarily store each message;
- send the complete message to the computer attached to a selected node or send it on to another node.

However, message switching may suffer 'clogging' if the buffer stores at nodes in the network have inadequate capacity to handle long messages.

19.7.3 Packet switching

This seeks to achieve a highly efficient use of the network by splitting up messages into short packets of data, each with a header containing information about the identity of the message, the sender and receiver, the sequence of the packets in a given message, the message priority and the identification of the last packet of a message.

Clearly, this involves additional data but the overall benefit of avoiding congestion from long messages represents a worth-while advantage.

19.8 The integrated services digital network (ISDN)

The economics of data communication and the uses of information technology are being profoundly influenced by the development of the world-wide 'Integrated Services Digital Network' (ISDN). ISDN provides a wide and growing range of 'dial up' modes of communication, including telephony, facsimile, videotex, audio and television conferencing (Chapter 18, section 18.2), and database transfer.

Thus, ISDN is essentially multi-purpose and is also capable of multipoint operation, e.g. for 'broadcast' distribution of messages.

The structure is based on the 64 kbit/sec bit rate for PCM digital speech transmission, which can be used in integral multiples of bit rate for wider bandwidth services, e.g. group-to-group video conferencing. Furthermore, since charges can be related to duration of use and bit rate, the ISDN can provide an economical service for the user. [6]

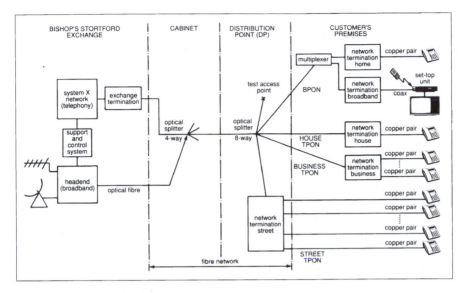

(a) passive optical fibre local network telephony (TPON)
and later broadband (BPON)

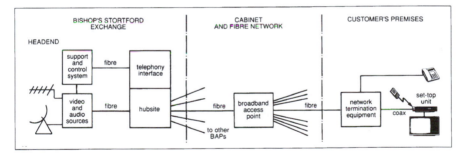

(b) active optical fibre star-switched broadband local network

Figure 19.3   Bishop's Stortford integrated services field trials on optical fibre (1990)

19.8.1 Milton Keynes New Town and Bishop's Stortford test-beds for ISDN

Milton Keynes New Town, Buckinghamshire, England has been described as ' . . . a town of the future, a showcase for new technology . . . one of the cornerstones of its success has been the forward thinking that has gone into the town's telecommunication and information technology'. [7]

By early 1988 Milton Keynes became the first urban area in the UK to be served entirely by a digital network with all the benefits of speedier connections and short-code dialling, better voice quality, and faster data transmission. The services available included tele-conferencing and digital data services such as:

- 'KiloStream' for electronic mail, credit verification, fast facsimile, and computer links to remote terminals;
- 'MegaStream' for high-speed data transmission at rates up to 2 Mbit/sec, e.g. for computer-to-computer data transfer, interconnection of digital private branch exchanges etc.;
- 'Packet Switch Stream', a national packet-switched network for non-voice communication at distance-independent tariffs, linked to similar networks in more than 50 other countries;
- 'SatStream' providing leased two-way circuits at rates between 56 kbit/sec and 2 Mbit/sec to Europe and North America via small satellite dishes.

The Milton Keynes cable network also provided sound and television services from the BBC and ITV, and satellite television channels for the home, together with a locally-produced information service using 'photo-video text' for the presentation of local news, shopping, housing, job and leisure information, and advertisements. Besides being available in the home, this service was presented on public viewdata terminals sited around the town.

The town of Bishop's Stortford in Hertfordshire was also the setting for a comprehensive field trial involving the provision of integrated services to domestic and business users. Two approaches were explored, with the aim of identifying the most economical method for meeting various customer requirements.

One approach used a 'passive' optical fibre local distribution network, i.e. one not involving active devices, to provide telephony and low-bit rate services at minimum cost while permitting up-grading to broadband services such as television in the future. Copper wire pairs or coaxial cable were used for the relatively short connections at the customers' premises.

Another approach used an 'active' switched-star network, similar to that in the Westminster cable-television trial (see below), adapted to include telephony; this method also used optical fibre up to the customers' premises.

19.8.2 The on-demand electronic video library service

In spite of a growing multiplicity of continuously running television channels by cable and satellite – each heavily loaded with material of only marginal interest to many viewers – there is an increasing public interest in the use of bought or hired video casettes offering full-length recordings of plays, films, documentary and other material. These, at least, offer the viewer the possibility of seeing what they want at a time of their own choosing – free from advertisements. This suggests that an on-demand video library service provided electronically over cable, a local area microwave radio network or – as latter became possible, via customer's telephone wire pairs to the local exchange – could offer an attractive viewing service with an almost unlimited choice of viewing material.

Payment for the service could be closely linked to the duration and quality of the material viewed, for example with higher rates for first-run films already acclaimed by the critics.

In 1986 British Telecom initiated a pioneering field trial of such a service over cable in the Westminster area of London, mainly to hotels. The viewer was presented on the TV receiver screen with a 'menu' of titles of items to view from which he made a choice by operation of a push-button keyboard. The video material offered was recorded on Philips 'LaserVision' discs, selected exclusively for each customer by mechanical selection at the video library centre on a machine reminiscent of the 'Juke-box' used to play audio records. The system was designed initially to accommodate up to 250 simultaneous sessions from some 10,000 customers, over the BT switched-star TV cable network. The switched-star network concept itself represents an important contribution to the economical distribution of television by cable – for his work on it Bill Ritchie of the BT Martlesham Research Laboratories was awarded the Martlesham Medal in 1988. [8]

Clearly, the future of the video library concept will depend on a much more sophisticated use of electronic techniques to store and access video programmes, e.g. on write/read computer discs that are now able to store some hours of video programme.

19.9 British Telecom trials of 'video on demand' over customers' telephone copper pairs

Far-reaching developments, based on advanced video digital coding techniques pioneered, amongst others, by ATT in America and British Telecom at its Martlesham, Suffolk Laboratories, enable TV broadcast-quality video signals to be transmitted over ordinary copper pairs used to provide telephone service to customers' premises. The digital compression achieved enables the bit rate to be reduced from the 100 Mbit/sec hitherto needed for good quality video to some 1.5 or 2 Mbit/sec. The technological consequences of this advance – termed 'ADSL' (Asymetric Digital Subscribers Loop) – are twofold:

- the digital signals can be transmitted over several kilometres of telephone copper pairs without loss of picture quality;
- an hour or more of video can be stored on computer discs or microchips, with random access on demand.

The video digital compression techniques are being studied with a view to world-wide standardisation by the International Organization for Standards (ISO) and the International Electrotechnical Commission (IEC); they have potential application not only to 'video on demand' but to TV broadcasting and other video services.

The BT field trial of 'video on demand' at Kesgrave, Suffolk in 1994 offered a modest choice of video material, but sufficed to demonstrate practicability. The proven 1.5 or 2 Mbit/sec capability of the copper pair local network clearly has potential for fast data, as well as video, services e.g. provided over the Internet.

References

1 Murray Laver, *Computers, Communications and Society* (Oxford University Press, 1975).
2 Murray Laver, *Information Technology: Agent of Change* (Cambridge University Press, 1989).
3 This account of the origin of Microsoft DOS is based on a discussion and interview with Bill Gates, Chief Executive of Microsoft, by the BBC in its *In Business* programme, June 1994.
4 C.P. Clare, *A Guide to Data Communications* (Castle House Publications Ltd, 1984).
5 'Information Networks' (*British Telecom Eng. Journal*, Vol. 9, Pt. 1, April 1990).
6 J.M. Griffiths, *ISDN Explained: World-Wide Network and Application Technology* (John Wiley and Sons, 1990).
7 'Building Tomorrow on Firm Foundations (Milton Keynes)' (*British Telecom World*, September 1989).
8 'On-Demand Interactive Video Services on the British Telecom Switched-Star Cable-TV Network' (*British Telecom Eng. Journal*, Vol. 4, No. 4, October 1986).

Chapter 20

Growth of the Internet and the World Wide Web

The Internet provides access to a global distribution of databanks containing stored information, and computers connected by telecommunication links to one another and to homes and offices throughout the world. At nodal points in the network are 'servers', specialised computers that can accept customer requests and control access to the appropriate databanks by providing automatic routing information. The selection of a desired databank is initiated by a coded address sent from a home or office computer.

This global network, to which perhaps half the population of the developed countries of the world now have access, constitutes the technology base for the World Wide Web, the information access, retrieval and processing service that has transformed – and continues to transform – the ways people live, work and play now and in the future. It has an important role in the management of national economies and the ways they are governed. As such it constitutes the 'nervous system and memory of mankind' – and is perhaps its greatest artefact.

As long ago as 1851 the poet Nathaniel Hawthorne wrote: [5]

By means of electricity the world of matter has become a great nerve, vibrating thousands of millions breathless points of time. The round globe is a very brain, instruct with intelligence!

20.1 The telecommunication developments that provide a technology base for the Internet

Although the Internet began to take shape around 1970 to meet USA defence requirements, the possibility for creating it on a world-wide scale for civilian use depended heavily on prior developments in telecommunications and computing, notably:

- the 'microchip' and 'CD' computer disc memory;
- stored-program software control of computers and exchange switching;
- the digital revolution;

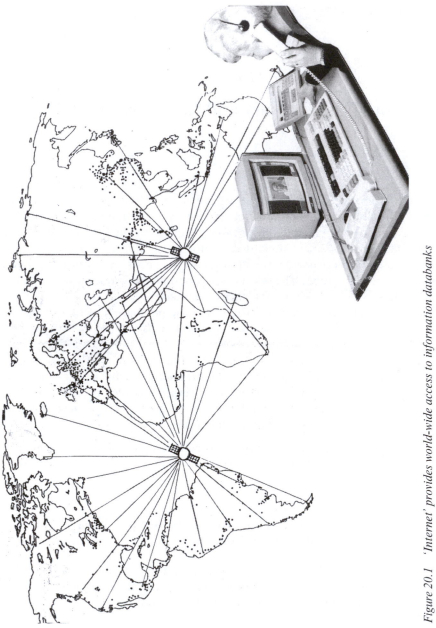

Figure 20.1 '*Internet*' *provides world-wide access to information databanks*

The network provides audio, facsimile, e-mail and visual communications over optical fibre cables and microwave links as well as the satellite links shown here

- the increased capacity and decreased cost per unit of bandwidth of telecommunication cable and satellite systems.

Certain areas of advanced telecommunication system technology proved to be of major importance for the creation of the Internet:

- the ISDN, the Integrated Services Digital Network which created a co-ordinated set of transmission standards for a wide range of services requiring different bit rates (see Chapter 19);
- the ATM, the Asynchronous Transfer Mode, pioneered by British Telecom, that facilitated the transmission of services requiring different bit rates on the same digital path;
- Packet Switching, pioneered by Donald Davies at the UK National Physics Laboratory, that enabled individual information messages to be split up into small or large packets depending on size, and transmitted via a common high-speed digital path. [1] [2]

20.2 The origins of the Internet and the World Wide Web

Any account of the origins of the Internet should begin with a reminder of the unique contribution of the British philosopher, scientist and mathematician Alan Turing whose work on the nature of information and its manipulation in the 1930s laid the foundation not only for computing but contributed to the cracking of the German Enigma code and the successful outcome of the Second World War (see Chapter 19). [3]

The story of the origins of the Internet during the last three decades is complex, spread over many individuals and organisations, academic, governmental and commercial, in the USA and Europe. The following outline is based on the compact version in the *Encyclopaedia Britannica Inc.* (1994–2001 edition) [4] and a detailed account *The Roads and Cross Roads of Internet History* by Gregory R. Gromov, now available as a 40-page web site. This presents a 'history with a human face', is at times amusing and always seeks to give a balanced presentation of the many personal and commercial rivalries involved in this remarkable story. [5]

The Internet began in 1969 as a research project called 'ARPANET', the US Government-supported Advanced Defense Project Agency Network, which searched for means to avoid disruption to telecommunication networks caused by enemy action. It was based on the development of a computer 'protocol' (software instructions) that enabled each computer to co-operate with others in the network; the project also included 'data packet switching' to enable fast operation of the linked network.

By 1972 a public demonstration had been given of the ARPANET system and its ability to restore communication in a defence telecommunication network subject to enemy action. However, doubts were raised whether any network could survive a nuclear war – not least because of the effects of the 'Electromagnetic Pulse' effect created by nuclear radiation that could destroy solid-state devices such as microchips.

In the 1970s groups of academic organisations in the USA, including the universities of California and Stanford, began developing means for linking their research computers in order better to share information and computing power. By 1986 this resulted in the 'US National Science Foundation Network' and links began to be set up with similar academic networks in the UK, Europe and Australia.

A major step forward was made in 1989 by Tim Berners-Lee and his associates at CERN (European Centre for Nuclear Research). Based in Geneva, CERN is the world's largest research centre, in which 40 nations and 370 scientific institutions co-operate in fundamental atomic particle research. This work involved the extensive use of computers to plan, execute and analyse the results of research. However, the computers used a variety of different operating standards which made inter-operation difficult if not impossible.

Tim Berners-Lee developed a 'Hyper Text Transfer Protocol' (http), initially used by CERN computers but which later laid a foundation for the World Wide Web, which standardised communication between users, servers and databanks. It enabled users seeking databank information to use a Web address identified by 'http' and send this from a personal computer with an ASCII keyboard via an appropriate server. (See Chapter 19, section 19.4.)

In November 1992 a 'Hyper-Text Project' was initiated by Tim Berners-Lee and R. Cailliou. It took searching the Web a stage further by creating a text-based Web 'browser', which enabled a user to search pages of information in a databank, and find addresses for links to other databanks.

20.3 The Internet had been born and the World Wide Web initiated!

Realising the commercial potential of what had been achieved at CERN, American computer manufacturers, notably 'Apple', followed by the giant IBM, began the development of the personal computer (PC), as did AMSTRAD in the UK.

From this point Internet development tempo in the USA expanded rapidly with several competing software companies, including Mosaic Communications and Netscape Communications, offering ever wider service facilities.

However, it was the advent of Bill Gates' 'Microsoft Corporation' that created widely accepted software for personal computers and a degree of standardisation throughout the world that was helpful in spreading Web usage (See Chapter 19, section 19.5). Nevertheless, Gates is reported as saying 'An Internet 'browser' is a trivial piece of software – there are least 30 companies that have written very credible Internet browsers – so that's nothing!'

From being primarily an 'information access' system in the 1970s the Internet/Web had evolved by 2000 into a global service of major importance to commerce, banking, insurance, newspapers/journalism, education and government itself, and it now provides person-to-person communication by electronic text/graphic mail 'e-mail' that is far quicker, and cheaper, than conventional letter mail.

Reliable statistics on the spread of the Internet and growth of Web usage are not available – the Web itself was created by 'organic' growth rather than planning by a centralising authority. However, 'data' traffic on the main telecommunication arteries of the world has more than doubled each year, and now exceeds by far conventional telephone traffic.

The future is likely to see a continuing growth, not only in the volume of traffic on the Internet but also in the range of services provided by the Web. The innovative advances in telecommunication technology that include high-capacity optical-fibre cables, fast opto-electronic switching and communication satellites of advanced design are well poised to provide for substantial growth of the Internet.

References

1 Donald Davies, NPL Pioneer of Packet Switching, Report of IEE Interview (*IEE Review*, July 1998).
2 Campbell-Kelly, 'Data Communications at NPL' (*Annals Hist. Computing*, Vol. 9, 3/4, 1988).
3 Turing Institute, UK. (www.turing.org.uk).
4 *Encyclopaedia Britannica* (1994–2001 edition). Search: World Wide Web, Internet, Computing-Servers.
5 Gregory R. Gromov, *The Roads and Crossroads of Internet History* (http://www.Internetvalley.com/intval.html).

Chapter 21

The development of the mobile radio service

21.1 The early history of mobile radio-communication

Mobile radio-communication may be described as communication with and between individuals who may be in land vehicles, ships, trains, or aircraft, via telephone, text, facsimile, e-mail or other services.

The history of mobile radio communication is as old as that of radio itself – much of Marconi's work in the early 1900s was stimulated by the need, especially for safety reasons, for communication with and between ships, a need which radio alone could satisfy (see Chapter 7). From this beginning grew from the 1920s onwards a world-wide development of coast radio stations for communication with ships, using the MF and HF regions of the radio spectrum and amplitude modulation (AM).

The Second World War gave a strong impetus to the development of VHF radio technology (30–300 MHz), both for communication in the fighting services and for radar. This led, from the 1950s, to wide-spread use of VHF for civilian land-mobile radio services such as the ambulance, police and fire services, and commercial users. Typically, these VHF systems used narrow-deviation frequency modulation, with 100-watt base stations and provided ranges up to some 50 miles. The invention of the transistor in the 1950s, and later the microchip, enabled the size and power requirements of the equipment to be substantially reduced and the reliability improved. Similar developments took place in maritime mobile radio, e.g. for ship-borne use, with international standardisation of the technical parameters and operational procedures via the International Telecommunication Union (ITU) and the International Radio Maritime Committee (CIRM). However, the expansion of these services and the growing demands on the VHF radio spectrum, which included broadcasting as well as mobile radio, made it necessary to seek new approaches to mobile radio communication, e.g. using frequencies in the UHF (300–3,000 MHz) range where more frequency space was available.

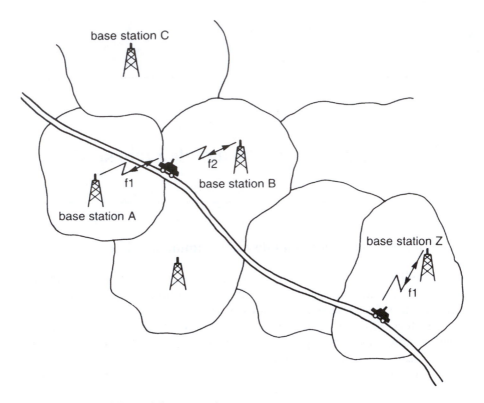

Figure 21.1   Cellular mobile UHF radio system

21.2 The UHF 'cellular' land-mobile radio system

However, a move to UHF brought its own problems. Shadow effects, e.g. of hills and tall buildings, are more pronounced than at VHF and the area effectively covered from each base-station is thus restricted, generally to a radius of a few tens of miles. Furthermore, because the wavelength is shorter, wave interference fading ('Rayleigh' fading) as seen from a moving vehicle, occurs faster than at VHF.

A major step forward was taken in the early 1970s when the American Bell System submitted a report to the US Federal Communication Commission proposing a 'cellular' mobile radio system using a multiplicity of marginally over-lapping base-station areas, each of some tens of miles radius. Except in the immediately adjacent base-station areas, this permitted re-use of frequencies in the various cells of the network and thus economy in spectrum utilisation [1] [2].

The base stations were connected into the public switched telephone network via fixed cable or microwave links; bearing in mind the multiplicity of mobile users in the many cells of the network and the time dispersion of their calls, this enabled a limited number of 'trunk' paths to provide a satisfactory grade of service.

An important technological development was the 'frequency synthesiser', based on a combination of digital techniques and phase-locked loops, which enables operating frequencies to be changed quickly and automatically as the mobile moves from cell to cell. [2]

As in other communication systems, cellular mobile radio has moved on from analogue (frequency modulation) to digital techniques offering important advantages, notably:

- low-bit rate encoding with forward error-correction which minimises the impairment of speech quality due to fading;
- encryption to ensure speech privacy.

International standardisation of the technical parameters (e.g. via the European Posts and Telecommunications Conference (CEPT) 'Global System for Mobile' (GSM) group) enables mobile users to move freely from country to country.

Looking to the future, there are interesting possibilities for using millimetric microwaves, e.g. in the absorption band around 60 GHz, to provide more capacity as demand grows. Although the cell size would be reduced, the mobile system might be integrated with the use of millimetric microwaves for providing local area fixed networks to customers.

21.3 Satellites and mobile radio-communication

For more than 15 years the international maritime satellite organisation IMARSAT has provided telephone communication with ships and aircraft. It has moved on to examine the possibility of world-wide voice, fax and data communication from ships and aircraft to land-mobile units via communication satellites.

Two types of satellite system have been proposed:

- 4 to 6 large satellites in an equatorial, 36,000 km height geo-stationary orbit;
- 12 smaller satellites at 10,000 km height in two orbital planes inclined at 47 degrees to a third, polar, plane.

Other orbital configurations have been proposed by various American companies. [3]

Although satellite systems could provide virtually world-wide coverage, including large land areas not covered by terrestrial cellular systems, there are difficulties. The large area covered by each satellite – which would include far more mobile users than each cell of a terrestrial cellular system – could give rise to problems of frequency allocation for large-scale utilisation.

21.4 Mobile radio communication at the turn of the century[1]

21.4.1 Background

Analogue cellular systems are commonly referred to as 'first generation' systems. They could carry speech only, were obsolete by 2000 and, by the end of 2001, should have been taken out of service. 'Second generation' systems are typified by the 'Global System for Mobile Communication', or GSM system, launched in the early 1990s. [4] This can carry speech and data at modest data rates, normally 9.6 kb/s but 14.4 kb/s under ideal conditions. Third generation systems are needed for multimedia communication.

The GSM cellular radiotelephone system is an internationally standardised mobile telecommunication system operating on digital principles able to carry both speech and data calls to and from the fixed, public switched telephone network (PSTN) and between mobile telephones. The mobile telephones can be used in any country that supports the system, subject to 'roaming' agreements between operators. This feature of international mobility, or roaming, was the main driving force of a nine-year European project, intended to offer manufacturers a standard to which they could make innovative products, assured of a very large, at least European, market. This large market produced a demand that drove down prices so that mobile telephony ceased to be a luxury and the GSM system is now employed in many other parts of the world as well as Europe. By 2001 it had become the most widespread of all mobile telephone systems with over 300 million users world-wide.

This has led to demand for more radio spectrum bandwidth to made available for mobile purposes and the third generation systems will require large numbers of wideband channels. Radio spectrum is allocated by government agencies according to a plan supervised by the International Telecommunications Union (ITU), a specialised agency of the United Nations. Over the years market demand and technological possibilities have increased the pressure on radio spectrum for mobile telephony, but there is never enough, which has led to the auctioning of spectrum for third generation systems operating in Europe, with prices (2000–2001) in the region of £ 1 billion/MHz.

21.4.2 Basic principles

A cellular radio telecommunication system has two outstanding characteristics. One is 'frequency re-use' and the other is 'handover' and these are explained in the following paragraphs.

To make efficient use of the scarce and costly resource of radio spectrum the radio frequency channels in a given allocation of spectrum are re-used many times over. The entire coverage area of a cellular system, typically a country, is divided into small radio areas called 'cells'. Each cell can use a small group of

---

[1] Contribution by Prof. David Cheeseman.

the total number of allocated radio channels and provided that the adjacent cells do not use any channels from the same group the problem of co-channel interference can be managed but not eliminated.

There are several ways of organising cells into groups (known as 'clusters') and a popular one uses a cluster size of 7 cells. To illustrate the idea, if 140 radio channels are allocated nation-wide and a 7-cell cluster is adopted, then each cell can use up to 140 divided by 7, or 20 radio channels. Each adjacent cluster then re-uses the same set of 140 channels in the same pattern, such that no adjacent cells employ the same channels. Cells vary in size from a few to several kilometres in radius to allow for variation in density of users and by this means the whole area of a country like the UK can be covered with a relatively small number of radio channels, yet provide service for millions of users.

This 're-use of frequency' is an essential feature of a cellular system. Of course, it leads to a demand for thousands of sites for the base stations, a situation that can cause environmentalists and lovers of the countryside to come into conflict with planners and telecommunication operators. This must be managed sensitively to accommodate the needs and views of both sides.

The other essential feature of a cellular system is 'handover'. Because the cells are small, mobile users can easily move out of coverage of the cell during a call; to maintain a connection it is necessary to set up a new radio channel in the new cell while clearing the old channel in the old cell. In principle, handover should achieve this without a noticeable break in transmission; in practice, because of the complexity of the system failures may occasionally occur.

A cellular system must have two basic features, a distributed network of radio transmitter/receiver, or base, stations each using a selection of the available radio channels and the ability to handover the connection from one to another as the mobile telephone moves throughout the coverage area.

The demand seen in fixed networks for Internet access can be met more satisfactorily by increasing the bandwidth available to the user. This is difficult in mobile networks, as in the GSM system, for example, the channel bandwidth is 200 kHz shared by the principle of time division by eight communication channels. Thus in effect each user has access to only 25 kHz which imposes a severe constraint on the data speed. By adopting a different modulation technique the GSM system has evolved to allow data speeds of up to 384 kb/s, but even this does not make effective provision for streaming or real-time video transmission.

To meet this requirement a completely new 'third generation' (3G) mobile system is under development which employs code division multiple access (CDMA) using typically 5 MHz radio channels. This offers enough bandwidth for a wide range of multi-media services. Each user has access to the whole 5MHz and is allocated a unique digital code which allows many users to occupy the same 5MHz of radio spectrum in a given cell. Equipment at the base station decodes the complex signal, separating out the individual users by their code. The cells themselves are allocated unique codes in a different range and the mobile equipment can distinguish between several different cells. This enables handover to be performed between cells, which in the case of this system may be necessary because of movement or because of the need, perhaps temporarily, for additional bandwidth. [5]

These codes have the effect of spreading the bandwidth of the transmitted signal so that the total transmitted energy occupies typically 5 MHz instead of 200 kHz. Because the energy is spread this results in a need for much lower levels of transmitted power for each user. However, a busy system with many operating channels has a power level that increases in proportion to the number of channels and this ultimately determines the amount of interference experienced by neighbouring cells which limits the capacity of the system as a whole.

The services for which this system is being developed will offer data rates of up to 2Mbit/s, far in excess of what can be offered today. The system is expected to come into operation in different parts of the world from the end of 2002 and will be the world's first 'universal mobile telecommunications system', known by the initials UMTS.

Hand-held mobile telephones are constrained by limited transmitter power, small keyboards and visual display screens (typically 3.5 cm × 4.5 cm) which restrict text to a few lines. The speed of data transmission is also low compared with computers connected to fixed links such as telephone lines and cables. Nevertheless they provide, in cost-acceptable form, simple communication facilities, much valued by users, such as:

- dial-up voice communication;
- display of text messages;
- 'paging' calls;
- basic e-mail.

By 2000, mobile radio systems in the UK had attracted some millions of customers via independent operators, including BT Cellnet, Orange, Vodafone and Virgin: similar large-scale development has taken place in the USA, Europe and Japan.

However, it must be acknowledged that there has been public concern about possible health risks, e.g. cancers or tumours, due to the electromagnetic radiation from the user equipment or base station transmitters. To date (2001) no firm medical evidence confirming or disproving this possibility has been brought forward.

The 3G development requires a much wider use of the radio-frequency spectrum than the present technology in order to provide increased communication bandwith for users. Use of the radio-frequency spectrum in this context by the competing telecommunication operators such as BT Cellnet, Vodafone and others has been decided by the UK Government auction of licences to operate – at costs of some billions of pounds a licence.

21.4.3 *Prospects for the next generation (3G) of mobile radio systems*

The technical considerations discussed earlier in this chapter have set the stage for a major new development in 'wireless' communication that will enable full-scale mobile access to the Internet and the World Wide Web, at present enjoyed by 'lap top' and 'desk top' computer users connected to the Internet via fixed links such as telephone lines and cables.

Co-ordination of this development is being carried out on the world scale by:

- the WAP (Wireless Application Protocol) forum created by Tim Berners-Lee; and
- W3C (the World Wide Web Consortium), dedicated to leading and advancing the development of the World Wide Web [6].

The higher costs involved in 3G, which will need more base stations, more complex mobile and base radio equipment, and higher operating costs (including those for licences to use the radio spectrum), make the economic future for 3G difficult to predict – and this in a world which will see the expanding growth of wide-bandwidth cable, optical fibre and millimetric-radio local area links which will offer ready access to the Internet at much lower cost.

The way of the 'innovator' is indeed hard!

References

1 Bell Laboratories Report to the Federal Communications Commission, *High Capacity Mobile System*, Technical Report, 1971. Also *Bell System Tech.Jl.* (Jan 1979).
2 'Frequency Synthesisers – a Review of Techniques', (*IEEE Tech. Trans.* COM 17, 1969).
3 D.B. Crosbie, 'The New Space Race – Satellite Communications' (*IEE Review*, May 1993).
4 M. Mouly and Pautet, *The GSM System for Mobile Communications*, M-B Cell and Sys., (4 Rue Elisee Reclus, F-91120 Plaiseau, France, 1992).
5 H. Holma and A. Toskala, *WCDMA for UMTS* (John Wiley & Sons).
6 'Wireless Application Protocol' WAP www.w3.org/TR/NOTE-WAP

Further Reading

Encarta Encyclopedia Article 'Cellular Radio Telephone' provides an account (to 1999) of Cellular Radio development in the USA.
www. http://encarta.msn.com Search: Cellular Radio Telephone.

Chapter 22

Telecommunications and the future

This review of innovators and innovation in telecommunications and broadcasting systems has shown how these have evolved from the primitive beginnings of electrical science in the 18th century to the comprehensive, powerful and worldwide communication systems of the 21st century, which are now indispensable to the social and business worlds and the processes of government. [1] [2]

The technology and system concepts that have been created would appear to provide well for existing communication service needs, both in types of service and by scale, and even to anticipate in some degree future needs. So what scope is there for innovation in the future?

The development of telecomunications had for many years been mainly 'technology led', that is the services provided were stimulated by the invention of devices and systems such as the microchip , coaxial and optical fibre cables , communication satellite systems, and computer-controlled electronic switching exchanges. However, there are on-going developments in micro- and opto-technology that may well stimulate the provision and wide availability of new services, as well as reducing the costs of providing established services.

The history of the development of the microchip in the last half of the 20th century demonstrated a dramatic and continuing increase in the number of transistor elements that can be accommodated on a single 1 cm square of silicon – doubling each year or so – and a substantial reduction in the cost of providing a memory or switching/processing function. In the early 1970s the number of such elements on a chip was about 2,000 with a line spacing of 6.5 microns and typical memory chips held 1,024 bits of information with 8-bit processing. By 1994, 35 million transistors per chip with a line spacing of 0.5 micron and 64-bit processing had been achieved. And by the end of the century 250 million-bit microchips became possible. Such increases in memory and processing power, together with improved video-coding methods will, for example, enable high-quality video programmes of an hour or more to be stored on a single microchip, opening the door for example to economically viable 'on demand' video viewing. The microchip technology initiated by Robert Noyce of Intel

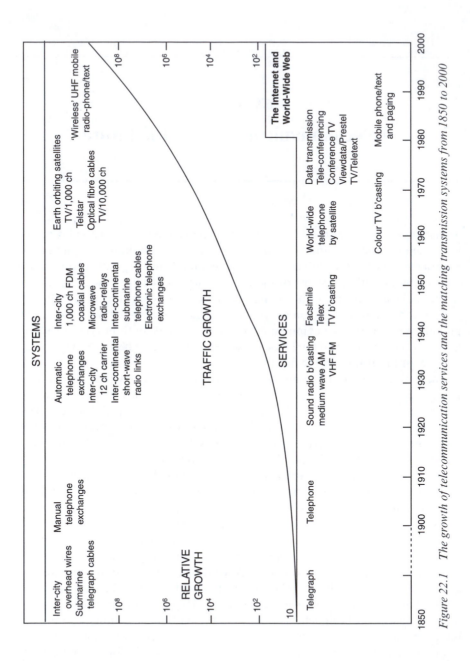

Figure 22.1   The growth of telecommunication services and the matching transmission systems from 1850 to 2000

and Jack Kilby of Texas Instruments in 1959 (see Chapter 12) has indeed borne fruit!

Nevertheless, it would appear that development in the near future will be mainly 'software led'; this will enable increasingly sophisticated programs to be created for the control of transmission, switching and network organisation, and facilitate the growth of existing – and provide new – services. The trend is thus towards a more 'intelligent network', one more 'user friendly' and better able to respond to, and even anticipate, customer requirements.

But we are now in an information revolution, christened 'cyberspace' by the media, in which digital highways on satellites and optical-fibre cables link computer-controlled switching nodes to create a world-wide network – the 'Internet' – with unlimited access from user's homes and offices to information databanks and direct person-to-person voice, facsimile, text and graphic communication (see Chapter 20). As bandwidth capacity on local and inter-city telecommunication links is increased higher-speed data transmission offering new services such as video become possible. This access has been extended to people on the move by the mobile cellular radio service that is itself poised to include data/video Internet access in the near future (see Chapter 21).

In the words of one commentator:

We are moving from a world where events are unrelated, and where heirarchical power structures predominate, into a world where the principal organizing structure is a network. In a network, everything – people, ideas and events – is linked, connective and inter-active. This fundamental change is the most significant aspect of what we are experiencing and will have the profoundest effect of all on our development as a species. [3]

22.1 Telecommunications and the integration of multi-media services and broadcasting

The evolution in recent years of integrated-service digital transmission and switching systems (ISDN), with an internationally standardised digital hierarchy accommodating a wide range of bit rates and services, provides a flexible framework within which voice, facsimile, e-mail, audio and visual teleconferencing, and video services – conveniently referred to as 'multi-media' services – can be provided on a common transmission path to customers' homes and offices. It could supplement the distribution of sound and television services currently provided by terrestrial and satellite radio transmitters and cable networks. It could also provide a forward-looking 'view on demand' television service, offering a valuable alternative to continuously-running television programmes.

The digital structure of the ISDN enables it to make use of advanced video coding techniques that reduce bit rate requirements for defined standards of picture quality, and facilitate video signal storage on microchips and CD discs.

The telephone copper pairs to customers' premises now offer potential via the

ADSL technology for good-quality video transmission and bandwidth for faster data/multi-media services. Since the telephone pairs are already available throughout the country this approach is enabling a start to be made in 2001 on the provision of multi-media services, including video.

The optical-fibre cable with its wide frequency bandwidth and convenient mechanical form offers virtually unlimited scope for the expansion of multi-media services into the local area, as well as intercity/international links.

Millimetric microwave radio and free-space optical links offer potential, as yet unexploited, for the provision of information bandwidths of 20MHz or more to customers' premises for local area distribution. The aerials/reflectors required are small, unobtrusive and quickly installed – they would avoid the environmental hazards of cable laying. [4] [5]

Existing local area coaxial cable systems, provided by independent operators, link customers' homes to central distribution points and provide 'broadcast' TV programmes and a limited choice of recorded video programmes, together with telephone service in some cases. Frequency-division multiplexing of the video channels is now being replaced by time-division digital multiplexing, and is moving towards the ISDN multi-service concept.

As optical-fibre replaces copper pairs and coaxial cables in the local network, providing even wider communication bandwidth, so the technology basis for an integrated network providing a wide range of multi-media services, attractive to both domestic and business users, becomes available.

Government legislation enforced by Oftel and created with the aim of promoting competition between independent cable system operators and the national telecommunication system provider British Telecom has tended to inhibit integration.

BT has not, for example, been allowed to offer video to its customers, or use millimetric microwave distribution in the local network. It has had to allow independent operators access to its copper-pair network to customers' homes, a process called 'unbundling of the local network'. This has required the installation of operators' equipment in BT local exchanges, a move which has involved considerable logistic and legal problems. However, the independent operators continue to rely on the national local exchanges, inter-city and inter-continental links in providing their services.

It is arguable that Oftel has inhibited the development of integrated system concepts and technologies that could have been major benefit to customers and good for export markets. It is to be hoped that the future will see a continuing move towards a well-planned and executed integrated national telecommunication system, so designed that it provides competition on a fair basis with and between independent service operators.

22.2 Television (video) conferencing

Person-to-person and group-to-group television conferencing facilities now available mark the beginning of a visual service with substantial potential for development and growth (see Chapter 18).

In particular may be noted the scope for innovation in large-screen, three-dimensional presentation with stereophonic sound for group-to-group conferencing to create a very real sense of 'you are here', virtual reality. This might involve the development of a multi-camera technique to avoid the lack of 'eye-to-eye' contact which occurs in a group situation with a single camera.

High-grade video tele-conferencing has important implications for the future – it could minimise the need for home-to-office and long-distance travel for business and governmental purposes, with significant economic, environmental and quality-of-life benefits.

22.3 The 'on-demand' video library service

An 'on-demand video library' service – in which the viewer has an immediate access over an integrated local distribution network to a wide range of recorded video programmes and video information sources in an 'electronic video library' – could provide what many would regard as a more acceptable form of video viewing than can be offered by a multiplicity of conventional continuously-running broadcast-television channels.

An electronic video library, analogous to a public book library, could contain an unlimited range of recorded video material, from old classic favourites to current-release films and today's football match. The choice of the video would be made via a 'user-friendly menu' presented on the customer's television receiver screen and giving an outline of contents, the code to select and the cost to view.

The latter is important – by making 'view-on-demand' a 'pay as you view service' it could be freed from the advertising that disfigures much of today's television viewing. And since a proportion of the revenue could go to the video programme producers, it could promote the making of higher quality programmes, capable of attracting larger viewing audiences.

It is to be noted that 'view-on-demand' requires a unique transmission channel between the video source and each customer's TV screen (similar to a telephone connection), but with the channel being occupied for perhaps more than an hour at a time. Since there could eventually be millions of customers in a given region, the service could not be provided by broadcast radio transmitters or satellites because of the excessive demand on radio spectrum frequency space that would be needed.

This points to a local 'electronic video library', akin to the local book library, as an early solution; even now many local book libraries offer computer video discs (CVD) for manual collection and rental. Video programme sources from further afield might well be accessed in the future as bandwidth available on national inter-city links and the Internet expands.

22.4 Domestic television viewing

22.4.1 Virtual reality

Progress in broadcast television domestic viewing is likely to continue the trend to higher definition and larger, flat-screen displays, making use of improved video coding and wider bandwidth transmission media.

A little-explored television viewing device is the video analogue of the personal headphone audio listening device, in the form of miniature screens built into the viewer's spectacles. Accompanying headphones could provide high-quality stereo sound. Since each eye can see a different picture it offers the possibility of realistic three-dimensional viewing.

The possibilities are interesting – such a device could give the ordinary domestic viewer the illusion of being in an unrestricted three-dimensional world, entirely cut off from his immediate surroundings. It also leads to the possibility of more specialised viewing devices offering 'virtual reality'. A viewer, equipped with a 3-D video head-set and hand-manipulated sensors, would be free to explore a picture databank at will.

The databank would need to contain the vast volume of picture information needed, for example, to view from all angles a building or to make journey in a town, or for a scientist to examine a DNA molecular system. Specialised applications of virtual reality viewing are beginning to appear, for example in the remote viewing at will from headquarters of work on a building construction site, or providing skilled guidance from a hospital for a remote surgical operation.

22.5 The electronic newspaper

The electronic newspaper of the future will probably use a flat display unit based on liquid-crystal and microchip techniques, thin and light enough to be comfortably hand-held, and linked by short-range radio to the telecommunication network for its sources of information. The displayed information could be transmitted overnight for reasons of economy, and up-dated as required during the day. An associated facsimile machine could provide hard-copy of selected articles.

Billing could be itemised and charged in terms of usage and items accessed. Such a system could offer 'newspapers' of virtually unlimited scope and variety, of local, national or overseas origin.

The social and environmental benefits could be substantial. A major advantage over the present-day large-scale use of paper for newsprint would be a slowing down of the rate of destruction of the world's forests.

The uneconomic use of man-power and the congestion of roads necessitated by today's cumbersome, slow and inefficient newspaper distribution system could be replaced by a speedy, energy-efficient electronic system.

22.6 An end to scientific journals?

Consideration is being given to the future of scientific journals in view of increasing production and distribution costs, the limited scale of the market and the need for speedier dissemination of information in a fast-moving scientific world. [4]

There seems no reason in principle why an electronic system, e.g. one similar to that outlined above for the electronic newspaper, should not be adapted for the preparation and distribution of scientific and similar specialised journals.

The advantages of so doing could be appreciable; not only could the costs and time involved in distribution be reduced, the editorial preparation of journals, including peer review and acceptance, could be speeded up. The immediate availability of hard copies of selected articles and data would be welcomed by scientific workers.

However, as the Internet continues to evolve it may be that this specialised service could be provided through the World Wide Web.

22.7 Opto-electronic and photonic switching: the transparent telecommunication network

Following the development of the optical fibre transmission systems and light amplifiers that are now part of the national and international telecommunication networks, several telecommunication laboratories, including British Telecom, are now exploring the possibilities of optical switching at exchanges, with a view to faster, more flexible and effective, and ultimately more economic, telecommunication services. This field is now wide open to innovation that could drastically change the structure and modes of operation of the network in the longer term future.

One approach is to employ optical switches using photons to replace the semiconductor devices using electrons in conventional exchange switches, but with essentially electronic control of the switching function – hence 'opto-electronic'. Another approach uses optical waves guided in a lithium niobate crystal which can be deviated from one guide to another by applied electrical control signals. This principle can be extended to light beams in space which can be deviated orthogonally to impinge on light-sensitive receptors on a two-dimensional matrix.

Multi-stage switching networks embodying these principles lend themselves to self-routing capabilities in which packet-type data with appropriate header addresses can find their own way through a complex network, setting up the switches on their own route as they proceed (see 'packet switching' in Chapter 19).

A radically advanced approach to optical switching leads to the concept of light-wavelength multiplexing based on highly coherent laser-light sources which can be marginally deviated in frequency to select one of a matrix of frequency selective light-sensitive receptors. Whilst this technique can be envisaged as applying initially to switching it leads to a more long-range concept of a completely 'transparent' network.

Proposers of this concept claim that there is sufficient bandwidth available in optical fibre networks, and the frequency stability in coherent light laser sources is potentially sufficient, for each customer in the national network to be allocated a unique frequency. Optical amplification could be used to overcome fibre and distribution losses. In such a system exchange switching would disappear and there would be complete freedom, within defined limits, to use the communication capability of a each connection thus established. [7]

22.8 The impact of telecommunications and IT on the future of mankind

But perhaps telecommunications is now moving into an era where technological innovation is relatively less important than innovation in the use of information technology for environmental improvement. It is a simple truism that it is no longer necessary to travel to communicate – there are now available the means to communicate at a distance, person-to-person, group-to-group and with or between computers and databanks in a variety of modes including aural, visual and permanent record (fax) forms. [1] [2]

A great deal of the work that now goes on in offices in large cities is concerned with the processing and exchange of information; to carry it out large numbers of commuters travel by train, bus and car, day by day, from and to their homes in the suburbs or other towns. Such travel is wasteful of time, involves the consumption of vast amounts of irreplaceable fossil fuels, and creates heavy peak demands on road and rail services leading to congestion and delays. At present information technology is being used on a relatively limited scale to replace travel – mainly by firms with existing offices in cities and factories in other locations. Long-distance inter-continental travel by jet aircraft for business purposes is wasteful of resources, time-consuming and destructive of the environment.

Tele-conferencing offers an economical and speedy alternative that further innovation will improve in communication effectiveness. Furthermore it offers a unique facility for conferring simultaneously between several users in widely different locations.

What is now needed to make more effective use of information technology is a co-operative and constructive approach by government, industry and those concerned with the environment to determine the benefits to the economy and the environment that could be achieved by the wider use of IT to minimise travel by enabling offices, factories and other centres of employment to be moved away from large cities and relocated where they were environmentally more acceptable.

If the benefits are both substantial and achievable – as they may well be – means for encouraging relocation and governmental support of a relocation planning policy for new offices and factories would be needed.

The innovative use of information technology could also have a beneficial effect on the rural economy, at present suffering from de-population and the loss of schools, shops and other facilities, by stimulating the development of small office and factory units in villages, linked to parent bodies elsewhere. There are possibilities too for 'working from home' or in a suitably equipped room in a village hall, i.e. a 'tele-cottage'.

A further benefit of the wider application of information technology to the geographical dispersal of office activity could well be to reduce the targets for terrorist bombs such has occurred in the City of London with its concentrated and massive office blocks in the city centre. An even more vivid example occurred on 11 September 2001 in New York. There would also be a degree of protection against damage from natural disasters such as Earthquakes, as experience in the USA has already demonstrated.

But, as noted earlier in this chapter, there is the overriding question of the impact that the world interactive network – the Internet – may have in the longer term on modes of thought, political processes and even the development of mankind itself.

It is clear that innovators and innovation will have a continuing and significant role to play in the future – perhaps with changing emphasis compared with the past – but nevertheless a role that could be of vital importance in determining where and how people live and work, and the quality of life they enjoy, in the years to come.

22.9 A national museum for electronic communication?

In recent years the writer has sought to develop public awareness, and that of professional bodies such as the IEE and the Royal Academy of Engineering, of the need for a new national museum for 'electronic communication', covering telecommunications, broadcasting and the Internet. [8] [9]

The pages of this book have demonstrated that the UK, from the Victorian scientists and mathematicians who laid the scientific foundations to the creators of the Internet, have made remarkable and vital contributions to the evolution of what is probably mankind's greatest artefact.

The primary aims of a national museum would be to:

- record for future generations the heritage of scientific and technological innovation that has created the world's telecommunication, broadcasting and information access (Internet) systems;

- present this story in a manner that would interest, inform and stimulate the imagination of today's younger generation and encourage them to make their own contributions in the future;
- present a view of the growing impact of electronic communication and IT on the way people will live and work in the future;
- present a 'show case for British industry' to illustrate its recent and most significant advances in telecommunications, broadcasting and information technology;
- at long-last honour the pioneers – many of them lone workers – whose endeavours made today's, and tomorrow's, worlds possible.

Given a decision to proceed at national level, major questions of organisation, funding and site location will need to be resolved. On the question of funding it should be remembered that the world communications industry, including the giants such as Microsoft and IBM, have made vast fortunes from the work of the pioneers – it is high time that this debt should be repaid.

References

1 Murray Laver, *Computers, Communications and Society* (Oxford University Press, 1975).
2 Murray Laver, *Information Technology: Agent of Change* (Cambridge University Press, 1989).
3 John May, 'The Shape of Things to Come', (*Sunday Telegraph Magazine*, September 1994).
4 T. Fowler, 'Mesh Networks for Broadband Access' (*IEE Review*, January 2001).
5 R. Detmer, 'A Ray of Light' (*IEE Review*, March 2001).
6 R. Smith, 'The End of Scientific Journals' (Letter, *British Medical Journal*, Vol. 304, March 1992).
7 J.E. Midwinter, 'Photonics in Switching: the Next Twenty Years of Optical Communications?', Chairman's Address, Electronics Division, (Institution of Electrical Engineers, London, October, 1990).
8 John Bray, 'Why the UK needs a National Museum for Electronic Communication' (*IEE News*, March 1999).
9 John Bray, 'A National Museum for Electronic Communication' (*Royal Acad. Eng. Jnl.* Vol.1, 2, October 1999).

Name Index

Subject Index